SpringerBriefs in Applied Sciences and Technology

Series Editor
Francis A. Kulacki
Department of Mechanical Engineering
University of Minnesota
Minneapolis
Minnesota
USA

For further volumes:
http://www.springer.com/series/8884

Yan Su • Jane H. Davidson

Modeling Approaches to Natural Convection in Porous Media

 Springer

Yan Su
University of Macau
Taipa
Macau

Jane H. Davidson
University of Minnesota
Minneapolis, Minnesota
USA

ISSN 2191-530X   ISSN 2191-5318 (electronic)
SpringerBriefs in Applied Sciences and Technology
ISBN 978-3-319-14236-4   ISBN 978-3-319-14237-1 (eBook)
DOI 10.1007/978-3-319-14237-1

Library of Congress Control Number: 2015930993

Springer Cham Heidelberg New York Dordrecht London

Springer is part of Springer Science+Business Media (www.springer.com)

Preface

Fluid flow and heat transfer in a porous medium are of interest in a number of engineering applications as well as in the environment. The primary purpose of this book monograph is to introduce modeling approaches for natural convection in porous media. These models are applicable to a wide variety of media, including sand, soil, randomly packed spheres or cylindrical tubes, and open cell metal foams, which have gained attention in recent years as potentially excellent candidates for meeting the high thermal dissipation demands in the electronics industry.

As an introduction to the topic of heat and mass transfer in porous media, Chap. 1 introduces the conventional defining parameters used to specify porous media and provides an overview of the governing equations and background material that set the stage for understanding modeling efforts. The local thermal equilibrium (LTE) and nonlocal thermal equilibrium (NLTE) approaches are introduced and compared.

Chapter 2 extends the theoretical presentation to consideration of the microscopic governing equations and volume-averaged macroscopic equations for flow and natural convection heat transfer. The theoretical development connects the microscopic drag and heat flux between solid and fluid phases in a porous medium through a recently developed geometry factor. Closure models are presented in a form applicable to a porous medium of arbitrary microscopic geometry.

Chapter 3 introduces numerical methods for simulation of natural convection in porous media, including the traditional finite difference based projection method and the nondimensional lattice Boltzmann method. The models are discussed in terms of the dimensionless governing parameters for natural convection. Mesh methods are presented for the finite difference and lattice Boltzmann numerical approaches.

Chapter 4 illustrates the application of the presented numerical methods to simulate the transient velocity and temperature fields and global heat transfer for in an adiabatic thin enclosure with an embedded heat sink. In this problem, the heat sink is a heat exchanger composed of multiple tubes. The example problem is an interesting application of the theory of porous medium to a practical engineering problem and illustrates the enormous power of treating a heat exchanger as a porous medium. Problem solutions are presented via porous medium model simulations by the projection method and the non-dimensional lattice Boltzmann method, and by direct numerical simulations with non-dimensional lattice Boltzmann method. Advantages and disadvantages of each numerical approach including comparison of the physical results and the CPU time are discussed. In addition, the use of the geometry factor to represent the porous medium is illustrated.

Contents

List of Abbreviations

a	the first Ergun constant
A	area, m^2
b	the second Ergun constant
$\mathbf{B}$	volumetric drag force, N/m^3
c	microscopic drag coefficient constant
c_h	microscopic heat transfer coefficient constant
c_p	specific heat at constant pressure, J/kg-K
C_D	microscopic drag coefficient, Eq. (2.14)
C_F	Forcheimer coefficient, Eq. (2.24)
d_H	hydraulic diameter, m
d	microscopic scale length scale, m
d_p	pore diameter, m
D	diameter of the enclosure, m
D2Q9	two-dimensional nine discrete velocity direction lattice mesh
Da	Darcy number, K/L^2
$\hat{\mathbf{e}}$	unit direction vector
$\mathbf{g}$	gravitational constant, kg-m/s^2
h	heat transfer coefficient, W/m^2-K
H	height of the enclosure, m
H_p	porosity layer height, m
k	thermal conductivity, W/m-K
k'	thermal dispersion conductivity, W/m-K
k_m	effective thermal conductivity, Eq. (2.31), W/m-K
K	permeability, m^2
ℓ	the thickness of microscopic structure of metal foam, m
L	macroscopic scale length scale, m
LBM	lattice Boltzmann method
LTE	local thermal equilibrium
M	constant $0 < M < 1$ for Eq. (2.31)
$\mathbf{n}$	solid fluid inter-surface direction vector in an REV, m
N	the grid number in the macroscopic length scale direction
NDLBM	non-dimensional lattice Boltzmann method

NLTE	non local thermal equilibrium
Nu_d	microscopic Nusselt number based on microscopic scale, hd/k_f
Nu_H	transient Nusselt number based on H, Eq. (4.14)
$\overline{Nu_H}$	time averaged Nu_H, Eq. (4.16)
Nu_m	Nusselt number based on porous medium $(k_f/k_m)Nu_H$, Eq. (4.15)
$\overline{Nu_m}$	time averaged Nu_m
p	volume averaged pressure, $\phi\hat{p}_f$, N/m^2
PPI	pore density, pores/inch
Pr	Prandtl number, v/α
q_{sf}	heat flux between the solid and fluid interface in per unit of REV, W/m^3
Ra_d	microscopic Rayleigh number, $\frac{g\beta\Delta T d^3}{v_f\alpha_f}$
Ra_H	macroscopic Rayleigh number, $\frac{g\beta\Delta T H^3}{v_f\alpha_f}$
Ra_m	porous medium Rayleigh number, $Ra_H Da k_f/k_m$
Re_d	microscopic Reynolds number, $|\hat{v}_f|d/v_f$
t	time, s
T	temperature, K
v	microscopic velocity, m/s
$\mathbf{v}$	Darcy velocity, $\phi\hat{\mathbf{v}}_f$, m/s
V	elementary volume in REV, m^3
	Greek symbols
α	thermal diffusivity, m^2/s
α'	thermal dispersion diffusivity, m^2/s
β	volumetric temperature expansion coefficient, K^{-1}
ϵ	thermal dispersivity
ϕ	porosity
ΔT	temperature scale, K
η	dimensionless geometry factor, $A_{fs}d/V_s$
μ	dynamic viscosity of the fluid, N-s/m^2
v	kinematic viscosity, m^2/s
ρ	density, kg/m^3
	Subscripts
f	fluid
h	heat transfer coefficient
m	porous medium
s	solid
	Superscripts
—	time average
$*$	dimensionless variable
$\widehat{}$	macroscopic variable
∞	far field

Chapter 1
Introduction of Fluid Flow and Heat Transfer in Porous Media

Abstract Background and basic concepts are introduced to define the defining parameters for study of porous media. The local thermal equilibrium (LTE) model and the nonlocal thermal equilibrium (NLTE) model of heat transfer in porous media are compared based on their assumptions and applicability. The key dimensionless governing parameters for a geometry, i.e., description of a porous medium are presented.

Keywords Local thermal equilibrium · Nonlocal thermal equilibrium · Representative elementary volume · Volume averaging

1.1 Specification of Porous Media

A porous medium is a material with a skeletal solid structure with void spaces. The void spaces allow fluid to pass through the medium. Numerous examples are present in nature including sand, soil, sponges, and porous and fractured rock, and in engineered materials such as metal and reticulated ceramic materials used for catalytic supports, structural members, filters, and extended heat transfer surfaces. Moreover, as will be illustrated in Chap. 4, practical engineering devices such as high surface area heat exchangers can be modeled as porous media. Porous media are described in multiple length scales. As illustrated by the sketch in Fig. 1.1, the macroscopic length scale is defined by the overall physical domain indicated by the length scale L. The microscopic length scale captures the detailed morphology and is indicated by d. For example, in the illustration, d is the diameter of an individual sphere. Usually the macroscopic length scale is far larger than the microscopic length scale $L \gg d$. A representative elementary volume (REV) is defined as a volume whose size ℓ_{REV} is large compared to the microscopic length scale d, but small compared to the macroscopic length scale L [4, 93], i.e., $d \ll \ell_{REV} \ll L$. The macroscopic variables are commonly defined by the volume average of the microscopic variables over the REV. It is assumed that the value of the macroscopic variables do not change when the averaging volume is larger than the REV [93].

The porosity is one measure of the ease with which fluid can flow through the structure. Porosity ranges from a low percentage in rock to about 50 % in sand, and over 90 % in some manmade materials such as reticulated foam or fiber board. The

Y. Su, J. H. Davidson, *Modeling Approaches to Natural Convection in Porous Media*, SpringerBriefs in Applied Sciences and Technology, DOI 10.1007/978-3-319-14237-1_1

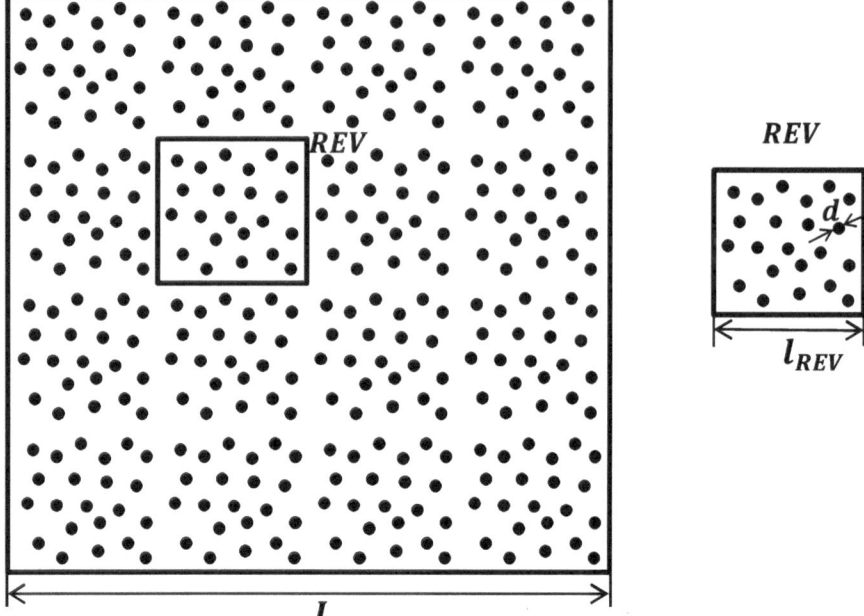

Fig. 1.1 Sketch of REV in a porous media packed by spheres

porosity of a porous medium is defined as the fraction of the volume of the fluid phase in an REV,

$$\phi = \frac{V_f}{V_s + V_f}.$$ (1.1)

A dimensionless geometry factor which connects the macroscopic and microscopic drag and heat flux between the solid and fluid phases in a porous medium was developed by Su et al. [74]. The geometry factor,

$$\eta = \frac{A_{fs} d}{V_s},$$ (1.2)

represents the distribution of the solid, which is specified solely by the shape of the solid. Unlike the widely used porosity based geometry factor, $(1 - \phi)$, which represents the fractional volume of the solid, η is a measure of how widely the solid is distributed in the fluid. Thus, for any porous medium of a given bulk porosity, a high value of η implies a large interface area between solid and fluid per unit volume, and $\eta(1 - \phi)$ represents the dimensionless interface area per unit volume. Using η, it is possible to model the fluid flow and heat transfer in a uniform format for various microscopic scales and geometries of the porous matrix [74]. For example, $\eta = 6$ for a packed bed of spheres, and $\eta = 4$ for loosely packed cylinders. The specific surface area expressed in terms of the geometry factor is,

$$ss = \frac{A_{fs}}{V_s + V_f} = \eta \frac{1 - \phi}{d} \qquad (1.3)$$

where d is the characteristic length of the solid fibers. Su et al. [74] show that the geometry factor for metal foams is accurately represented by,

$$\eta = \frac{0.025 \times 6d \left[\pi d_p d + 3\sqrt{3}(d_p + d)^2/4 - \pi d_p^2/2 \right]}{(1 - \phi)(d + d_p)^4}, \qquad (1.4)$$

where d_p is the characteristic dimension of the pore. The pore size, conventionally given in pores per inch, PPI, is,

$$\text{PPI} \approx \frac{0.0254}{d + d_p}. \qquad (1.5)$$

1.2 Porous Media Models

Fluid flow and heat transfer in a porous medium are of interest in a variety of engineering applications as well as in the environment [10, 83]. Models of heat transfer and flow in porous media have been applied to a wide variety of media, e.g., sand, soil, randomly packed spheres [3, 10] or cylindrical tubes [69, 33], and open cell metal foams, which have gained attention in recent years as potentially excellent candidates for meeting the high thermal dissipation demands in the electronics industry [37, 60].

1.2.1 Models for Fluid Flow

1.2.1.1 Darcy Equation

Based on experimental data, Darcy proposed a linear relationship between the volume averaged fluid velocity $\mathbf{v}$ and the pressure difference across the porous media Δp,

$$\mathbf{v} = k \frac{\Delta p}{\Delta x} \qquad (1.6)$$

In this expression, k is the hydraulic conductivity and $\frac{\Delta p}{\Delta x}$ is the pressure gradient. The hydraulic conductivity describes the ease with which fluid can flow through the pore spaces. As discussed by Lage [40], the Darcy's equation (1.6) is valid for incompressible and isothermal creeping flows.

1.2.1.2 Hazen–Darcy Equation

Darcy's experiments were performed with a constant temperature single fluid, so the Darcy equation (1.6) does not include the effect of fluid viscosity μ. Although Hazen [27] observed the influence of fluid temperature on the Darcy equation, it was not until Muskat [50] that fluid viscosity μ was added into the Darcy equation. By introducing the specific permeability $K = k\mu$, the Darcy's equation is re-expressed in a viscosity-dependent form as,

$$\mathbf{v} = \left(\frac{K}{\mu}\right)\frac{\Delta p}{\Delta x}. \tag{1.7}$$

The specific permeability K is a hydraulic parameter, assumed to be independent of fluid properties. Equation (1.7) is the Hazen–Darcy equation, which is also called Darcy's Law.

1.2.1.3 Hazen–Dupuit–Darcy Equation

From the analysis of the steady open channel flow, a quadratic Hazen–Dupuit–Darcy equation is derived from a balance of the gravity force and the shear resistance [40],

$$0 = \frac{\partial p}{\partial x} - \frac{\mu}{K}\mathbf{v} - C\rho\mathbf{v}^2. \tag{1.8}$$

The second and third terms on the right hand side of Eq. (1.8) represent the lumped viscous drag and the lumped form drag, respectively, where C is the form parameter. Equation (1.8) is valid for one-dimensional, steady-flow in porous media. Ward [87] suggests replacing the form parameter C by $C_F/K^{1/2}$, where the inertial coefficient C_F is a function of porous parameters,

$$0 = \frac{\partial p}{\partial x} - \frac{\mu}{K}\mathbf{v} - \frac{C_F}{K^{1/2}}\rho\mathbf{v}^2. \tag{1.9}$$

Based on the study of Ergun [16], the permeability K and the inertial coefficient C_F are given by,

$$K = \frac{\phi^3 d_p^2}{a(1 - \phi)^2}, \tag{1.10}$$

$$C_F = \frac{b}{\sqrt{a}\phi^{3/2}}, \tag{1.11}$$

where a and b are Ergun constants ($a = 150$, $b = 1.75$ for forced convection; and $a = 215$, $b = 1.92$ for natural convection). The permeability Reynolds number, Re_K can be used to determine flow regime,

$$Re_K = \frac{\rho\mathbf{v}K^{1/2}}{\mu} = \frac{1}{C_F}\frac{\rho C K}{\mu}\mathbf{v}, \tag{1.12}$$

because it is equivalent to the ratio of form and viscous forces,

$$\frac{D_C}{D_\mu} = \frac{C\rho\mathbf{v}^2}{\frac{\mu}{K}\mathbf{v}} = \frac{\rho C K}{\mu}\mathbf{v}. \tag{1.13}$$

1.2.1.4 Brinkman–Hazen–Dupuit–Darcy Equation

Brinkman noticed that the fluid viscous shear stress effect can be negligible compared to the viscous drag for low permeability. By adding the shear stress term (Laplacian term) to the three-dimensional form of Eq. (1.9), the Brinkman–Hazen–Dupuit–Darcy equation is obtained,

$$0 = -\nabla p + \mu\nabla^2\mathbf{v} - \frac{\mu}{K}\phi\mathbf{v} + \frac{C_F}{K^{1/2}}\rho\phi^2|\mathbf{v}|\mathbf{v}. \tag{1.14}$$

As discussed by Lage [40], the major contribution of Brinkman is the recognition that the fluid can transmit viscous shear force independently of the viscous drag. Both K and C_F can vary widely over the range of porous media relevant to engineering, e.g., packed beds, stacked screens, metal foams, and aerogels. Experimental data for metal foams [37, 85] show that when PPI $= 10$ (pore density, or pores per inch) $K \approx 1.03$ m^2, and $C_F \approx 0.07$ for $\phi = 0.88$, and $K \approx 1.6$ m^2 and $C_F \approx 0.06$ for $\phi = 0.92$. This variation implies that the first of the Ergun constants [16] in Brinkman's model [7] varies with porosity.

1.2.1.5 Recent Models Based on Volume Averaging

The volume averaging method of Whitaker [88] relates the volume average of a spatial derivative to the spatial derivative of the volume average, and makes the transformation from microscopic equations to macroscopic equations possible. Vafai [82] and Hsu [29] simplified the volume averaging method of Whitaker [88] making it widely useful for engineering applications. The governing equation for conservation of momentum of Hsu and Cheng [29] can be presented in form,

$$\frac{\partial\mathbf{v}}{\partial t} + \bar{\nabla}\cdot\left(\frac{\mathbf{vv}}{\phi}\right) = -\bar{\nabla}p + \mu_f\bar{\nabla}^2\mathbf{v} - \frac{\mu}{K}\rho_f\phi\mathbf{v} - \frac{C_F}{K^{1/2}}\rho_f\phi^2\mathbf{v}|\mathbf{v}|$$
$$- C_{12}\rho_f\phi^{3/2}\mathbf{v}|\mathbf{v}|^{1/2}, \tag{1.15}$$

where C_{12} a coefficient due to the drag of the microscopic flow passing a sphere and is associated with the fluid viscous shear stress effect [29]. The above momentum equation (1.15) gives the full map of the forces acting on an REV. We can see that the fluid viscous shear stress effect should become negligible compared to the viscous drag when $\frac{L}{d}$ is large. Hence, it is easy to understand why early experimental study ignored the viscous shear stress effect. When the buoyancy force is considered, the momentum equation is,

$$\frac{\partial \mathbf{v}}{\partial t} + \hat{\nabla} \cdot \left(\frac{\mathbf{vv}}{\phi} \right) = -\hat{\nabla}p + \mu_f \hat{\nabla}^2 \mathbf{v} - \frac{\mu}{K} \rho_f \phi \mathbf{v} - \frac{C_F}{K^{1/2}} \rho_f \phi^2 \mathbf{v} |\mathbf{v}|$$

$$- C_{12} \rho_f \phi^{3/2} \mathbf{v} |\mathbf{v}|^{1/2} - \phi \beta \mathbf{g} (\hat{T}_f - T_{ref}). \qquad (1.16)$$

The porous medium model of Hsu and Cheng [29] provides the relation between bulk porosity and the Ergun constants for packed spheres, and implies an inherent relation between macroscopic drag, the microscopic drag coefficient, and the microscopic geometry. However, this relation can not be applied to other small scale geometries such as metal foam. Hsu [31] developed a drag model based on an effective hydraulic diameter, $d_H = d\phi/(1 - \phi)$, because the interference among solid particles is important, and the closure coefficients depend strongly on the porosity. Nakayama [54] and Braga's [6] analysis of turbulence in a packed bed of spheres is based on the microscopic length scale, d, and channels or pipes based on the hydraulic diameter d_H. However the hydraulic diameter does not decouple the effects of porosity and the small scale structure of the solid on drag and heat flux per unit volume of a porous medium. The geometry factor, η of Eq. (1.2) developed by Su et al. [74] provides the link between the macroscopic and microscopic drag and heat flux through the interface of the solid and fluid phases in a porous medium. The details of the model of Su et al. [74] based on the geometry factor η will be discussed in Chap. 2.

1.2.2   Models for Heat Transfer

Based on the volume averaging method, there are two heat transfer models: the one-equation model based on the local thermal equilibrium (LTE) assumption and the two-equation model based on the nonlocal thermal equilibrium (NLTE) assumption. The two models will be discussed separately based on their different assumptions and application limitations.

1.2.2.1   LTE Model

By distinguishing the gradient operators in the microscopic and macroscopic co-ordinates, a simple form of the volume average was presented by Hsu [29]. The macroscopic equations for forced convection of an incompressible flow in a variable porosity medium were obtained by a volume averaging of the microscopic conservation equations over an REV,

$$\left(\rho C_p \right)_m \frac{\partial \hat{T}}{\partial t} + \bar{\nabla} \cdot \left[\mathbf{v} \hat{T} \right] = \alpha_f \left(\frac{k_m}{k_f} + \frac{k'}{k_f} \right) \hat{\nabla}^2 \hat{T}. \qquad (1.17)$$

Hsu [30] suggests methods to calculate k_m/k_f, while the method to obtain k'/k_f is a topic of ongoing study. Based on the assumption of local thermal equilibrium (LTE), the one-equation model of the energy equation (1.17) can greatly save computational time. As discussed by Vafai and Amiri [80], the LTE assumption implies that the difference between the local fluid and solid temperatures is negligible or is much smaller than the temperature difference for the global system. This assumption can introduce inaccuracies in the model of energy transport.

1.2.2.2 NLTE Model

The two-equation model, shown in Eqs. (1.18) and (1.19) by Riaz [64] and Vafai [80], reduces the inaccuracy of LTE assumption when the flow velocity is extremely high or when the heat source in one of the two phases is much higher than in the other phase [65].

$$\phi \left(\rho C_p \right)_f \left[\frac{\partial \hat{T}_f}{\partial t} + \hat{\nabla} \cdot (\mathbf{v} \hat{T}_f) \right] = \phi \alpha_f \left(\frac{k_m}{k_f} + \frac{k'}{k_f} \right) \hat{\nabla}^2 \hat{T}_f + h_{sf}(\hat{T}_s - \hat{T}_f) A_{sf},$$

$$(1.18)$$

$$(1 - \phi) \left(\rho C_p \right)_s \frac{\partial \bar{T}_s}{\partial t} = (1 - \phi) \alpha_s \hat{\nabla}^2 \hat{T}_s - h_{sf}(\hat{T}_s - \hat{T}_f) A_{sf}. \qquad (1.19)$$

The two-equation model is called the NLTE model because it allows the macroscopic solid temperature to differ from the macroscopic fluid temperature.

A heat transfer coefficient between the solid and fluid interface h_{sf} is required to close the model. A great deal of uncertainty and inconsistency are found between the reported experimental results for h_{sf}. An experimental correlation for forced convection of bed packed with spheres is given by Wakao and Kaguei [86] in Eq. (1.20), which is valid for $100 < Re_d < 8500$,

$$Nu_{sf} = \frac{h_{sf}d}{k_f} = 2 + 1.1 Re_d^{0.6} Pr_f^{1/3}, \qquad (1.20)$$

Re_d is the Reynolds number based on the Darcy velocity and the particle diameter defined as,

$$Re_d = \frac{|\mathbf{v}|d}{\nu_f}. \qquad (1.21)$$

For unconsolidated porous media packed by separated spheres, a correlation based on the asymptotic solution for heat transfer of Falkner–Skan forced convection over a passing sphere was obtained by Nakayama et al. [55],

$$Nu_{sf} = 1.47 Re_d^{1/2} Pr_f^{1/3}. \qquad (1.22)$$

This correlation agrees well with Eq. (1.20) over a wide range of the Reynolds number from 10 to 10^5.

For consolidated porous media such as foams, Fu et al. [20] correlated experimental data for airflow through mullite and cordierite ceramics in form,

$$Nu_{sf} = n_0 Re_{d_p},\tag{1.23}$$

where $n_0 = 0.275$ for mullite with $\phi = 0.916$ and $n_0 = 0.099$ for cordierite with $\phi = 0.742$. Kamiuto and Yee [35] obtained a correlation for unsteady flow through metal foam based on the pore diameter,

$$Nu_{sf} = 0.124 \left(\frac{3\pi\phi}{4(1-\phi)} \right)^{0.605} \left(Re_{d_p} Pr_f \right),\tag{1.24}$$

where Re_{d_p} is the Reynolds number based on the Darcy velocity and the pore diameter,

$$Re_{d_p} = \frac{|\mathbf{v}| d_p}{\nu_f}.\tag{1.25}$$

Nakayama et al. [55] also provided a correlation based on a volume averaged analysis of the macroscopic energy equation,

$$Nu_{sf} = 0.07 \left(\frac{\phi}{(1-\phi)} \right)^{2/3} \left(Re_{d_p} Pr_f \right).\tag{1.26}$$

Recently, Xu et al. [90], measured air flow through low porosity metal foam with $0.16 \leq \phi \leq 0.38$, and obtained,

$$Nu_{sf} = (0.993\phi^2 - 0.245\phi + 0.0165) Re_d^{0.8} Pr_f^{1/3}.\tag{1.27}$$

To facilitate the application of the correlations, a uniform correlation is necessary and will be discussed in Chap. 2.

Chapter 2
A Uniform Theoretical Model for Fluid Flow and Heat Transfer in Porous Media

Abstract To comprehensively study fluid flow and heat transfer in porous media, a uniform model that is valid for both a regular microscopic geometry, e.g., spheres and cylinders, and irregular microscopic geometries, such as metal foams or ceramic, is needed to connect the macroscopic and microscopic flow and heat transfer. In this chapter, the microscopic governing equations and volume averaged macroscopic governing equations for flow and natural convection heat transfer are presented. As defined in Chap. 1, the dimensionless geometry factor connects the macroscopic and microscopic drag and heat flux between the solid and fluid phases in a porous medium. Based on this geometry factor, the closure models are presented based on the microscopic drag coefficients and heat transfer correlations in a uniform form for porous media of arbitrary microscopic geometry. Also, relationships between the microscopic drag coefficients and permeability, Forchheimer coefficient, and Ergun constants are presented.

Keywords Drag coefficients · Geometry factor · Heat transfer coefficients

2.1 Microscopic Governing Equations

The microscopic continuity equation for an incompressible flow in porous media is,

$$\nabla \cdot \mathbf{v}_f = 0. \tag{2.1}$$

The microscopic momentum equation is,

$$\rho_f \left[\frac{\partial \mathbf{v}_f}{\partial t} + \nabla \cdot \left(\mathbf{v}_f \mathbf{v}_f \right) \right] = -\nabla p_f + \mu_f \nabla^2 \mathbf{v}_f + \left(\rho_f - \rho_{f\infty} \right) \mathbf{g}. \tag{2.2}$$

The microscopic energy equations for the fluid and solid phases are,

$$\left(\rho C_p \right)_f \left[\frac{\partial T_f}{\partial t} + \nabla \cdot \left(\mathbf{v}_f T_f \right) \right] = \nabla \cdot \left(k_f \nabla T_f \right), \tag{2.3}$$

and

$$\left(\rho C_p \right)_s \frac{\partial T_s}{\partial t} = \nabla \cdot \left(k_s \nabla T_s \right) + q, \tag{2.4}$$

© The Authors 2015

Y. Su, J. H. Davidson, *Modeling Approaches to Natural Convection in Porous Media*, SpringerBriefs in Applied Sciences and Technology, DOI 10.1007/978-3-319-14237-1_2

where q is a heat source in the solid phase. When the interface is treated as a surface with zero thickness, the interface conditions that guarantee continuity of temperature and heat flux between fluid and solid phases are,

$$T_f = T_s \quad on \ A_{fs}, \tag{2.5}$$

and

$$\mathbf{n}_{fs} \cdot k_f \nabla T_f = \mathbf{n}_{fs} \cdot k_s \nabla T_s \quad on \ A_{fs}. \tag{2.6}$$

2.2 Macroscopic Governing Equations

Whitaker [88] applies a volume averaging method to derive the macroscopic governing equations in porous media. The volume average of a spatial derivative is related to the spatial derivative of the volume average. Recently, Vafai [82] and Hsu [29] simplified the volume averaging method of Whitaker [88] and provided closure models for the thermal dispersion.

By distinguishing the gradient operators in the microscopic and macroscopic coordinates, a simple form of the volume average is presented by Hsu and Cheng [29] for the fluid (f) and solid (s) phases in a porous medium. An intrinsic phase average of a quantity associated with the fluid phase is defined as,

$$\hat{W}_f = \frac{1}{V_f} \int_{V_f} W_f \, dV, \tag{2.7}$$

where V_f is the volume occupied by the fluid phase in V and $V_f + V_s = V$.

By introducing the method of volume averaging as given in Eq. (2.7) for the velocity and temperature deviations in the porous medium, we obtain the macroscopic continuity momentum and energy equations, as discussed in [29].

The macroscopic continuity equation is,

$$\hat{\nabla} \cdot \mathbf{v} = 0, \tag{2.8}$$

where $\mathbf{v} = \phi \hat{\mathbf{v}}_f$ is the Darcy velocity vector.

The macroscopic momentum equation with the Boussinesq assumption is,

$$\rho_f \left[\frac{\partial \mathbf{v}}{\partial t} + \hat{\nabla} \cdot \left(\frac{\mathbf{v}\mathbf{v}}{\phi} \right) \right] = -\hat{\nabla} p + \mu_f \hat{\nabla}^2 \mathbf{v} + \mathbf{B} - \rho_f \phi \beta \left(\hat{T}_f - \hat{T}_\infty \right) \mathbf{g}, \tag{2.9}$$

where

$$\mathbf{B} = -\frac{1}{V} \int_{A_{fs}} p_f \, d\mathbf{S} + \frac{\mu_f}{V} \int_{A_{fs}} (\nabla \mathbf{v}_f) \cdot d\mathbf{S}, \tag{2.10}$$

and

$$p = \phi \hat{p}_f. \tag{2.11}$$

On the right-hand side of Eq. (2.9), the second term is the viscous shear in the fluid. The third term, represented by **B**, is the drag force per unit volume between solid and fluid surfaces, i.e., the pressure and viscous drag force per unit volume of the porous media.

The macroscopic energy equations for fluid and solid are respectively,

$$\phi \left(\rho C_p\right)_f \left[\frac{\partial \hat{T}_f}{\partial t} + \hat{\nabla} \cdot \left(\hat{\mathbf{v}}_f \hat{T}_f\right)\right] = \phi \hat{\nabla} \cdot \left[\left(k_f + k'\right) \hat{\nabla} \hat{T}_f\right] + q_{sf}, \qquad (2.12)$$

$$(1 - \phi) \left(\rho C_p\right)_s \frac{\partial \hat{T}_s}{\partial t} = (1 - \phi) \left[\hat{\nabla} \cdot \left(k_s \hat{\nabla} \hat{T}_s\right)\right] - q_{sf} + q. \qquad (2.13)$$

2.3 Closure Models for Macroscopic Equations

2.3.1 Closure Model for Drag

With a matched asymptotic expansion similar to that used for tubes [69, 71] and spheres [29, 54], the drag coefficient, C_d, for an arbitrary microscopic geometry with the microscopic length scale d can be expressed as,

$$C_d = c_{d_0} + c_{d_1} Re_d^{-1} + c_{d_2} Re_d^{-1/2} + O(Re_d^{-3/2}), \qquad (2.14)$$

where,

$$Re_d = \frac{|\hat{\mathbf{v}}_f| d}{\nu_f}, \qquad (2.15)$$

and c_{d_0}, c_{d_1}, and c_{d_2} are constants. The zeroth order term is a correction associated with the inertial effect, the -1 order term is the Stokes drag, the $-1/2$ order term is due to the skin friction, and the $-3/2$ order term is a negligible higher order term. Hence, the drag per unit volume of the porous medium can be expressed as,

$$\mathbf{B} = \frac{Drag_{fs}}{V_s + V_f} = -\frac{\frac{1}{2}\rho_f |\hat{\mathbf{v}}_f| \hat{\mathbf{v}}_f A_{fs} C_d}{V_s/(1 - \phi)}$$

$$= -(1 - \phi)\eta \frac{\rho_f}{2} \frac{v_f^2}{d^3} \left[c_{d_0} Re_d^2 + c_{d_1} Re_d^1 + c_{d_2} Re_d^{3/2}\right] \hat{\mathbf{e}}_f, \qquad (2.16)$$

where, $\hat{\mathbf{e}}_f$ is the unit vector in the direction of the macroscopic velocity, i.e., the Darcy velocity, $\hat{\mathbf{v}}_f/|\hat{\mathbf{v}}_f| = \mathbf{v}/|\mathbf{v}|$. From Eq. (2.16), it can be seen that the geometry factor, η, presented in Eq. (1.2), and the bulk porosity, ϕ, connect the macroscopic drag force and the microscopic drag coefficient for arbitrary microscopic geometry.

2.3.2 Relation to the Darcy–Brinkman Model

The drag model presented in Eq. (2.16) is a uniform format, which is appropriate for arbitrary microscopic geometries. The frequently used Darcy–Brinkman's model presented in Eq. (1.14) in Chap. 1 is a special case of Eq. (2.16). The two models are equivalent when the skin friction term of Eq. (2.14) is negligible or when the fluid viscous shear stress effect is negligible compared to the viscous drag, i.e., when $\frac{L}{d}$ is large.

For viscous drag in packed bed of spheres, the drag force based on Brinkman's model [7] is,

$$\nabla \hat{p}_f = -\left[\frac{\mu_f}{K}\mathbf{v} + \frac{C_F \rho_f}{\sqrt{K}}\mathbf{v}|\mathbf{v}|\right]. \tag{2.17}$$

By multiplying Eq. (2.17) by ϕ, the drag force per unit volume is,

$$\mathbf{B} = \nabla p = \nabla \phi \hat{p}_f = -\phi\left[\frac{\mu_f}{K}\mathbf{v} + \frac{C_F \rho_f}{\sqrt{K}}\mathbf{v}|\mathbf{v}|\right], \tag{2.18}$$

which can be expressed as,

$$\mathbf{B} = -\left[\frac{\mu_f}{K}\phi^2\left(\frac{v_f}{d}\right)Re_d + \frac{C_F \rho_f}{\sqrt{K}}\phi^3\left(\frac{v_f}{d}\right)^2 Re_d^2\right]\hat{\mathbf{e}}_f. \tag{2.19}$$

Comparing Eqs. (2.16) and (2.19), based on the coefficients of zero and -1 order, one obtains,

$$\frac{\mu_f}{K}\phi^2\left(\frac{v_f}{d}\right) = (1-\phi)\eta\frac{\rho_f}{2}\left(\frac{v_f}{d}\right)^2\frac{1}{d}c_{d_1}, \tag{2.20}$$

and

$$\frac{C_F \rho_f}{\sqrt{K}}\phi^3\left(\frac{v_f}{d}\right)^2 = (1-\phi)\eta\frac{\rho_f}{2}\left(\frac{v_f}{d}\right)^2\frac{1}{d}c_{d_0}, \tag{2.21}$$

respectively. From Eqs. (2.20) and (2.21), the permeability and the Forchheimer coefficient can be expressed in terms of the drag coefficient and geometry factor for an arbitrary structured porous medium as,

$$K = \frac{\phi^2}{(1-\phi)\eta}\frac{2d^2}{c_{d_1}}, \quad \text{and} \quad C_F = \frac{\sqrt{(1-\phi)\eta}}{\phi^2}\frac{c_0}{\sqrt{2c_{d_1}}}. \tag{2.22}$$

Thus the Darcy number can be recast as,

$$Da = \frac{K}{L^2} = \frac{\phi^2}{(1-\phi)\eta}\left(\frac{d}{L}\right)^2\frac{2}{c_{d_1}}. \tag{2.23}$$

Equations (2.22) and (2.23) show the effects of the microscopic geometry factor, porosity, and the microscopic drag coefficients on permeability, Forchheimer coefficient, and Darcy number. From Ergun's experimental study [16] of a bed of packed spheres, permeability and the Forchheimer coefficient are related to the porosity by,

$$K = \frac{\phi^3 d^2}{a(1-\phi)^2}, \quad \text{and} \quad C_F = \frac{b}{\sqrt{a}\phi^{3/2}}. \tag{2.24}$$

Thus the Ergun constants can be expressed as,

$$a = \frac{\phi}{(1-\phi)} \frac{\eta}{2} c_{d_1}, \quad \text{and} \quad b = \frac{\eta}{2} c_{d_0}. \tag{2.25}$$

Based on prior studies [29] for a packed bed of spheres, $\eta = d(\pi d^2/4)/(\pi d^3/24) = 6$, $c_{d_0} = 0.4/4 = 0.1$, $c_{d_1} = 24/4 = 6$, and $c_{d_2} = 6/4 = 1.5$. For loosely packed cylinders [69], $\eta = d(\pi d)/(\pi d^2/4) = 4$, $c_{d_0} = 1.18/\pi$, $c_{d_1} = 6.8/\pi$, and $c_{d_2} = 1.96/\pi$. Thus the Ergun constants, a and b, differ a great deal depending on the structure of the porous matrix. From Eq. (2.25), it is seen that b is independent of porosity while a is dependent on it. Thus we see why experimental correlation constants for a vary more than b when the porosity changes, as discussed in [7, 10] and [16].

From Eq. (2.16), a normalized drag force is derived,

$$|\mathbf{B}| \frac{d^3}{\frac{1}{2}\rho_f v_f^2 (1-\phi)} = \eta \left[c_0 Re_d^2 + c_1 Re_d^1 + c_2 Re_d^{3/2} \right] = \eta C_D Re_d^2. \tag{2.26}$$

This dimensionless quantity is independent of porosity and linear in η.

2.3.3   Closure Models for Heat Transfer in Porous Media

2.3.3.1   LTE Model

The governing energy equation of the LTE model is identical to Eq. (1.17),

$$\left(\rho C_p\right)_m \frac{\partial \hat{T}}{\partial t} + \bar{\nabla} \cdot \left[v \hat{T} \right] = \alpha_f \left(\frac{k_m}{k_f} + \frac{k'}{k_f} \right) \hat{\nabla}^2 \hat{T} + (1-\phi)q. \tag{2.27}$$

The effective thermal conductivity of the porous medium is expressed as,

$$\frac{k_m}{k_f} = \phi + (1-\phi)\frac{k_s}{k_f}, \tag{2.28}$$

and the effective heat capacity is expressed as,

$$\frac{(\rho c_p)_m}{(\rho c_p)_f} = \phi + (1-\phi)\frac{(\rho c_p)_s}{(\rho c_p)_f}. \tag{2.29}$$

Hence, the ratio of thermal diffusivities is,

$$\frac{\alpha_m}{\alpha_f} = \frac{\phi + (1 - \phi)\frac{k_s}{k_f}}{(\phi + (1 - \phi)\frac{(\rho c_p)_s}{(\rho c_p)_f})}. \tag{2.30}$$

The above mixture model has some limitations. For complex geometries like metal foams, the effective thermal conductivity may be represented by [5],

$$\frac{k_m}{k_f} = M\left[\phi + (1 - \phi)\frac{k_s}{k_f}\right] + \frac{(1 - M)}{\left(\phi + \frac{1-\phi}{k_s/k_f}\right)}, \tag{2.31}$$

where $M = 0.33$ based on recent experiments of natural convection in an open cell metal foam [85]. (In [5], $M = 0.35$ was recommended for one-dimensional conduction.)

Furthermore, when the convective effects of the fluid motion are introduced, thermal dispersion becomes important. Hsu [52] extended his earlier work of interfacial heat transfer for pure conduction [29] to incorporate the effect of forced convection for both low and high Reynolds number flows.

The solution of the energy equation requires a closure model for thermal dispersion. For forced convection in packed cylinders [29],

$$\frac{\alpha'}{\alpha_f} = \frac{k'}{k_f} = \begin{cases} \varepsilon\frac{1-\phi}{\phi}Re_d Pr_f, & \text{if } Re_d \gg 10, \\ \varepsilon\frac{1-\phi}{\phi^2}(Re_d Pr_f)^2, & \text{if } Re_d \ll 10, \end{cases} \tag{2.32}$$

where α' is the thermal dispersion diffusivity, k' is the the thermal dispersion conductivity, and ε is the thermal dispersivity. Similarity analysis has demonstrated a closure model for dispersion in natural convection [69, 71]. For natural convection, the local velocity is estimated to be $\sqrt{g\beta\Delta T d}$, and thus the local Reynolds number is,

$$Re_d \sim \frac{\sqrt{g\beta\Delta T d d}}{\nu_f} = \sqrt{\frac{Ra_d}{Pr_f}}. \tag{2.33}$$

Equations (2.32) and (2.33) are combined to yield,

$$\frac{\alpha'}{\alpha_f} = \frac{k'}{k_f} = \begin{cases} \varepsilon\frac{1-\phi}{\phi}(Ra_d Pr_f)^{1/2}, & \text{if } \frac{Ra_d}{Pr_f} \geq 100, \\ \varepsilon\frac{1-\phi}{\phi^2}(Ra_d Pr_f), & \text{if } \frac{Ra_d}{Pr_f} < 100. \end{cases} \tag{2.34}$$

The quantities α' and k' are tensors in anisotropic porous medium, but it is common to treat them as isotropic scalars. For forced convection of water and air through heated packed channels and cylindrical packed tubes, $\varepsilon \sim 0.04$ [29]. In our previous work [69, 74], we found that the enhancement of thermal conductivity due to thermal dispersion with value 0.04 is 1–3% for both cylinders and metal foams, and thus heat transfer enhancement due to thermal dispersion is insignificant.

2.3.3.2 NLTE Model

The governing energy equations of the NLTE model are,

$$\phi \left(\rho C_p\right)_f \left[\frac{\partial \hat{T}_f}{\partial t} + \hat{\nabla} \cdot \left(\hat{\mathbf{v}}_f \hat{T}_f\right)\right] = \phi \hat{\nabla} \cdot \left[\left(k_f + k'\right) \hat{\nabla} \hat{T}_f\right] + q_{sf}, \qquad (2.35)$$

$$(1 - \phi) \left(\rho C_p\right)_s \frac{\partial \hat{T}_s}{\partial t} = (1 - \phi) \hat{\nabla} \cdot \left(k_s \hat{\nabla} \hat{T}_s\right) - q_{sf} + q, \qquad (2.36)$$

where q_{sf} is the heat flux from solid to fluid phase and q is a heat source in solid phase, such as electrical heaters, or as will be shown in Chap. 4 for tube bundle heat exchangers, treated as porous medium.

The heat flux between the solid and fluid phases per unit volume of porous media q_{sf} can be expressed as,

$$q_{sf} = \frac{A_{sf} h_{sf} (\hat{T}_s - \hat{T}_f)}{\frac{V_s}{(1-\phi)}} = h_{sf} (\hat{T}_s - \hat{T}_f) ss = (1 - \phi) \eta \frac{1}{d} h_{sf} (\hat{T}_s - \hat{T}_f). \qquad (2.37)$$

In a manner similar to that for microscopic drag, an asymptotic expansion for the microscopic heat transfer coefficient is,

$$h_{sf} A_{fs} (T_s - T_f) = \oint_{REV} k_f \left(\frac{\partial T}{\partial \mathbf{n}}\right)_{sf} dA_{fs}. \qquad (2.38)$$

With Eq. (2.38), the microscopic Nusselt number can be expressed as,

$$Nu_d = \frac{h_{sf} d}{k_f} = c_{ho} + c_{h1} Re_d^{c_{h2}} Pr_f^{c_{h3}}. \qquad (2.39)$$

For packed spheres, $c_{ho} = 2$, $c_{h1} = 0.6$, $c_{h2} = 1/2$, and $c_{h3} = 1/3$ [88], and for loosely packed cylinders, $c_{ho} = 0.3$, $c_{h1} = 0.62$, $c_{h2} = 1/2$, and $c_{h3} = 1/3$ [71]. For the saturated metal foam investigated by Wade [85], we observe that there are many wedge shapes, or nearly so, in the microscopic structures. We therefore compare the wedge flow heat transfer coefficients (with $c_{ho} = 0.3$, $c_{h1} = 0.51$, $c_{h2} = 0.5$, and $c_{h3} = 1/3$ [89]) to widely used correlations for fibrous metal foams [97], and find that they show reasonably good agreement over a wide range of Re_d.

Substituting Eq. (2.39) into Eq. (2.37), the heat flux per unit volume is,

$$q_{sf} = (1 - \phi) \eta \left[c_{ho} + c_{h1} Re_d^{c_{h2}} Pr_f^{c_{h3}}\right] \frac{k_f}{d^2} (\hat{T}_s - \hat{T}_f), \qquad (2.40)$$

and the normalized heat flux is,

$$\frac{q_{sf} d^2}{k_f (\hat{T}_s - \hat{T}_f)(1 - \phi)} = \eta \left[c_{ho} + c_{h1} Re_d^{c_{h2}} Pr_f^{c_{h3}}\right] = \eta Nu_d, \qquad (2.41)$$

which, like the microscopic drag coefficients, are independent of ϕ and linear in η. The grouping ηNu_d demonstrates the importance of the microscopic interface on heat transfer.

Chapter 3
Numerical Methods

Abstract Macroscopic, microscopic, and mesoscopic length scale-based dimension-less governing parameters for natural convection in porous media are introduced. Dimensionless governing parameters at three length scales are presented. Dimensionless governing equations with closure models are derived based on the geometry factor discussed in Chap. 2. Both the conventional finite difference based projection method and the nondimensional lattice Boltzmann method (NDLBM) are presented. Mesh methods for these two numerical schemes are discussed.

Keywords Lattice Boltzmann method · Mesoscopic length scale · Nondimensional lattice Boltzmann method · Projection method

3.1 Dimensionless Governing Parameters

Before discussion of numerical methods, we introduce the dimensionless governing parameters for natural convection in a porous medium. As discussed in Chap. 1, the microscopic length scale of a porous medium is represented by d, and the macroscopic length scale is represented by L. For computational simulations an additional length scale based on the mesh size ℓ, which is also called the mesoscopic length scale in lattice Boltzmann method (LBM) simulations, is introduced. For a uniformly distributed two-dimensional mesh $\ell = \Delta x = \Delta y$.

Including the two governing parameters porosity ϕ and the geometry factor η defined in Chap. 1, the governing parameters for natural convection in porous media are the Rayleigh numbers and the Prandtl numbers. The microscopic Rayleigh number is defined based on the microscopic length scale d,

$$Ra_d = \frac{g\beta\Delta T d^3}{\nu_f \alpha_f} \tag{3.1}$$

and the macroscopic Rayleigh number is defined based the macroscopic length scale L,

$$Ra_L = \frac{g\beta\Delta T L^3}{\nu_f \alpha_f} \tag{3.2}$$

© The Authors 2015   17
Y. Su, J. H. Davidson, *Modeling Approaches to Natural Convection in Porous Media*,
SpringerBriefs in Applied Sciences and Technology, DOI 10.1007/978-3-319-14237-1_3

The mescoscopic Rayleigh number based on the mescoscopic length scale, i.e., the mesh size ℓ, is

$$Ra_\ell = \frac{g\beta\Delta T \ell^3}{\nu_f \alpha_f}. \tag{3.3}$$

The Prandtl number based on the fluid phase is

$$Pr_f = \frac{\nu_f}{\alpha_f}. \tag{3.4}$$

The Prandtl number for numerical simulations is determined by the effective thermal diffusivity,

$$Pr_\ell = Pr_f \frac{\alpha_f}{\alpha_m}, \tag{3.5}$$

where the effective thermal diffusivity is given in Eq. (2.30).

3.2 Dimensionless Governing Equations

To obtain the dimensionless governing equations for natural convection in a porous medium, the length scale is L, the temperature scale is ΔT, and the velocity scale is $U = \sqrt{g\beta L \Delta T}$. The governing dimensionless macroscopic equations for NLTE model are,

$$\hat{\nabla}^* \cdot \mathbf{v}^* = 0, \tag{3.6}$$

$$\frac{\partial^* \mathbf{v}^*}{\partial^* t^*} + \frac{1}{\phi}\left(\mathbf{v}^* \hat{\nabla}^*\right)\mathbf{v}^* = -\hat{\nabla}^* p^* + \sqrt{\frac{Pr_f}{Ra_L}}\hat{\nabla}^{*2}\mathbf{v}^* + \mathbf{B}^* - \phi\hat{T}_f^* \hat{\mathbf{e}}_g, \tag{3.7}$$

$$\phi\frac{\partial^* \hat{T}_f^*}{\partial^* t^*} + \hat{\nabla}^* \cdot \left(\mathbf{v}^* \hat{T}_f^*\right) = \sqrt{\frac{1}{Pr_f Ra_f}}\hat{\nabla}^* \cdot \left[\phi\left(1 + \frac{k'}{k_f}\right)\hat{\nabla}^* \hat{T}_f^*\right] + q_{sf}^*, \tag{3.8}$$

$$(1 - \phi)\frac{(\rho C_p)_s}{(\rho C_p)_f}\frac{\partial^* \hat{T}_s^*}{\partial^* t^*} = \sqrt{\frac{1}{Pr_f Ra_f}}\hat{\nabla}^* \cdot \left[(1 - \phi)\left(\frac{k_s}{k_f}\right)\hat{\nabla}^* \hat{T}_s^*\right] - q_{sf}^*, \tag{3.9}$$

where $\hat{\mathbf{e}}_g$ is the unit vector in the gravity direction.

Based on the closure model discussed in Chap. 2, the dimensionless drag force is,

$$\mathbf{B}^* = -\frac{\eta}{2}\frac{(1-\phi)}{\phi}\left(\frac{L}{d}\right)\left[\frac{1}{\phi}c_0|\mathbf{v}^*|\mathbf{v}^* + \left(\frac{Pr_f}{Ra_L}\right)^{\frac{1}{2}}\left(\frac{L}{d}\right)c_1\mathbf{v}^*\right.$$
$$\left. + \left(\frac{Pr_f}{Ra_L}\right)^{\frac{1}{4}}\left(\frac{1}{\phi}\frac{L}{d}\right)^{\frac{1}{2}}c_2|\mathbf{v}^*|^{\frac{1}{2}}\mathbf{v}^*\right], \tag{3.10}$$

and the dimensionless heat flux at the solid–fluid interface is,

$$q_{sf}^* = \sqrt{\frac{1}{Pr_f Ra_L}}(1-\phi)\eta\left(\frac{L}{d}\right)^2\left[c_{h0} + c_{h1}Re_d^{Ch2}Pr_f^{Ch3}\right](\hat{T}_s^* - \hat{T}_f^*). \tag{3.11}$$

Eqs. (3.6)–(3.9) along with Eqs. (3.10) and (3.11) can be solved using either of two standard approaches, local thermal equilibrium (LTE) [69] and nonlocal thermal equilibrium (NLTE) models [71].

LTE models can be applied when the heat flux through the interface is balanced in both solid and fluid phases. The time scale of the conduction in the solid phase approaches the time scale of convection in the fluid phase when $k_s/k_f \sim LNu_d/(dN)$, where N is the grid number in the macroscopic length scale direction. Thus, even for high Ra_H, the LTE model can be applied with enough grid numbers. The dimensionless macroscopic energy equation becomes,

$$\left[\phi + (1-\phi)\frac{(\rho C_p)_s}{(\rho C_p)_f}\right]\frac{\partial^*\hat{T}^*}{\partial^*t^*} + \hat{\nabla}^* \cdot (\mathbf{v}^*\hat{T}^*) = \sqrt{\frac{1}{Pr_f Ra_L}}\hat{\nabla}^* \cdot \left[\left(\frac{k_m}{k_f} + \frac{k'}{k_f}\right)\hat{\nabla}^*\hat{T}^*\right] + \dot{q}^*.$$
$$\tag{3.12}$$

3.3   Review of Numerical Methods

Conventionally there are two numerical approaches for computational fluid flow and heat transfer: one is the discrete approach, and the other is the continuum approach [51]. The discrete approach starts from the discrete Hamilton's equations based on the microscopic scale. The discrete approach requires molecular dynamics simulations. The continuum approach starts from the continuum Navier–Stokes equations based on the macroscopic scale. The continuum approaches include finite difference [69], finite volume [2], finite element [39], and spectral methods [1], and are based on the volume averaged macroscopic equations.

Recently, the LBM based on a mesoscopic scale which is between the microscopic and macroscopic scales has been proposed to increase the simulation speed by using more computer memory compared to the conventional continuum methods and also to better handle complex geometry domains and multi-phase flows [11, 28]. Several studies based on an adjusted Darcy velocity ($\frac{v^*}{a+\sqrt{a^2+bv^*}}$, where a and b are coefficients determined by porosity, permeability, and space and time steps) show the possible applications of the LBM on heat transfer in porous media [21, 63, 67]; however, the physical explanation of the adjusted coefficients are not clear.

3.4 Projection Method

3.4.1 Introduction of Projection Method

The fractional step method was introduced by Chorin [12] as an effective means of solving the incompressible Navier–Stokes equations. The fractional step methods for incompressible flows are often referred to in the literature as projection methods [24]. The fraction step methods are classified into three classes, namely the pressure-correction method, the velocity-correction method, and the consistent splitting method [24]. Typically the projection method operates as a two-stage fractional time step scheme [59]. For each numerical time-step, there are four steps as follows:

1. The transport equations for mass and momentum are solved using a suitable advection method, which is called the predictor step.
2. An initial projection is implemented to enforce the mid-time-step velocity field as divergence free.
3. The corrector part of the algorithm is then progressed.
4. The divergence restraint on the velocity field is enforced to update to a new time step.

The advantage of the projection method is that the computations of the velocity and the pressure fields are decoupled.

3.4.2 Projection Method with Stagged Meshes

As discussed by Reyret and Taylor [59], an implicit second-order Adams–Bashforth scheme of projection method with staggered meshes is introduced for incompressible flow simulations. This method was applied in natural convection in porous medium by [69]. The momentum equation with buoyancy is discretized into Eqs. (3.13) and (3.14). The energy equations are discretized into the Eqs. (3.15) and (3.16).

$$\frac{\widetilde{\mathbf{v}}^* - \mathbf{v}^{*n}}{\Delta\tau} + \frac{1}{\phi}\frac{3\left[(\mathbf{v}^*\hat{\nabla}^*)\mathbf{v}^*\right]^n - \left[(\mathbf{v}^*\hat{\nabla}^*)\mathbf{v}^*\right]^{n-1}}{2} = \sqrt{\frac{Pr_f}{Ra_L}}\frac{\hat{\nabla}^{*2}\mathbf{v}^{*n}}{2} + \mathbf{B}^{*n+1} - \phi\hat{T}_f^{n+1}\mathbf{n_g},$$

$$(3.13)$$

$$\frac{\mathbf{v}^{*n+1} - \widetilde{\mathbf{v}}^*}{\Delta\tau} = -\hat{\nabla}^* p^{n+1/2} + \sqrt{\frac{Pr_f}{Ra_L}}\frac{\hat{\nabla}^{*2}\mathbf{v}^{*n+1}}{2},$$

$$(3.14)$$

$$\phi \frac{\hat{T}_f^{*(n+1)} - \hat{T}_f^{*n}}{\Delta\tau} + \left[\mathbf{v}^*\hat{\nabla}^*\hat{T}_f^*\right]^{n+1} = \phi\sqrt{\frac{1}{Pr_f Ra_L}}\left[1 + \frac{\alpha'}{\alpha_f}\right]\hat{\nabla}^{*2}\hat{T}_f^{*(n+1)} + q_{sf}^{*(n+1)},$$

$$(3.15)$$

$$\left[(1-\phi)\frac{(\rho C_p)_s}{(\rho C_p)_f}\right]\frac{\hat{T}_s^{*(n+1)} - \hat{T}_s^{*n}}{\Delta\tau} = (1-\phi)\sqrt{\frac{1}{Pr_f Ra_L}}\left(\frac{k_s}{k_f}\right)\hat{\nabla}^{*2}\hat{T}_s^{*(n+1)}$$

$$+ (1-\phi)q^{*(n+1)} - q_{sf}^{*(n+1)}, \qquad (3.16)$$

where, $\tilde{\mathbf{v}}^*$ is a visual velocity used to discretize the two steps for the calculation of the velocity.

Based on Eq. (3.17), if we take the divergence of the Eq. (3.14), we get the Poisson equation for the pressure $p^{n+1/2}$ as Eq. (3.18),

$$\nabla \cdot \mathbf{v}^{*n+1} = 0, \qquad (3.17)$$

$$\nabla^2 p^{n+1/2} = \frac{1}{\Delta t}\nabla \cdot \tilde{\mathbf{v}}^*. \qquad (3.18)$$

At a wall boundary, the pressure can be expressed,

$$\left(\frac{\partial p}{\partial N}\right)_\Gamma^{n+1/2} = 0. \qquad (3.19)$$

The numerical algorithm is the following. First, Eq. (3.13) is solved explicitly for $\hat{\mathbf{v}}^*$ and then the coupled Eqs. (3.14), (3.15), (3.16) and (3.18) are coupled and are solved by Gauss–Seidel iteration.

Temperature is calculated at the same point as the pressure, while velocity is not. The size of the matrices used to calculate the pressure and velocity field are different. For example in the rectangular domain shown in Fig. 3.1, the matrix size of the temperature and pressure is $M \times N$, while the matrix size for the x- and y-components of velocity u and v are $(M + 1) \times N$ and $M \times (N + 1)$, respectively. The staggered meshes are useful to assure continuity of velocity and pressure across any interfaces between the porous and nonporous zones, such as are encountered in metal foams with superposed fluid or for the example presented in Chap. 4, in which a heat exchanger immersed in a enclosure is treated as a porous medium.

3.5 Lattice Boltzmann Method

3.5.1 Introduction to Lattice Boltzmann Method

Recently, the LBM has been proposed to increase the simulation speed for complex geometry domains and multiphase flows [11, 28]. Unlike the projection method,

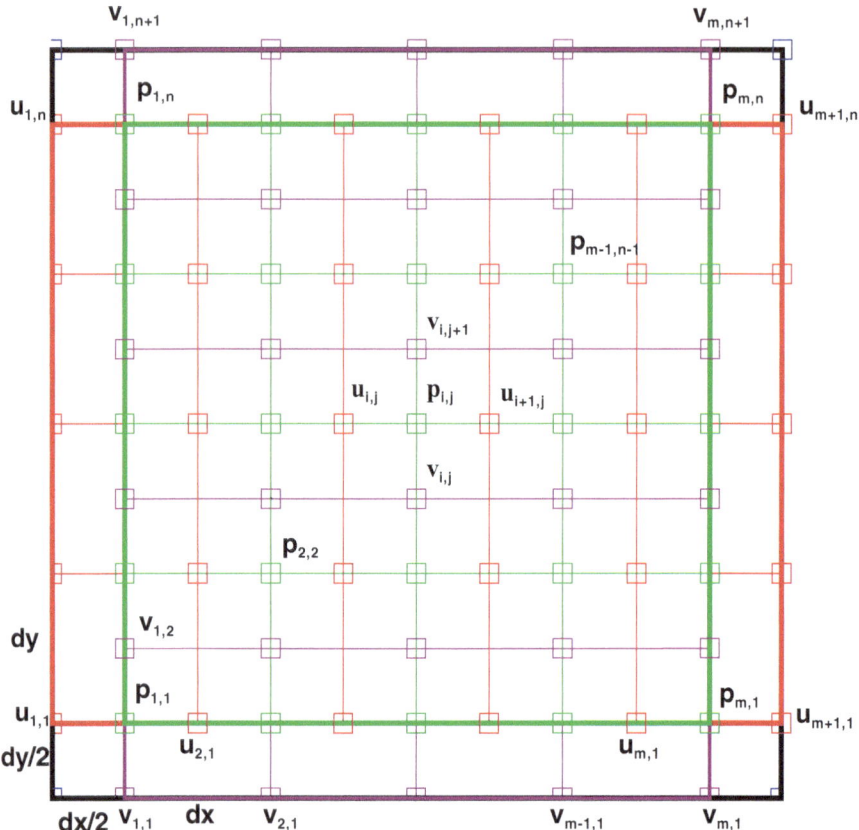

Fig. 3.1 Two-dimensional staggered meshes

which solves the conservation macroscopic governing equations, the LBM models the fluid as consisting of particle groups, and such particle groups perform collision and streaming processes on a discrete lattice mesh [51, 52]. The LBM has been applied for natural and mixed convection by introducing a body force term into the momentum equation [25, 49]. Also, the conjugate heat transfer of a solid and fluid was solved easily with LBM by a modification of the relaxation time in the energy equation [34, 77]. However, the conventional LBM based on real units cannot show explicitly the effects of the multiple length scale dimensionless governing parameters and the size of mesh on computational accuracy and speed. Hence the conventional LBM is usually limited to relatively low Reynolds and Rayleigh numbers. In order to overcome the above limitations of the conventional LBM, a novel nondimensional lattice Boltzmann method (NDLBM) was developed by Su et al. [75] to simulate conjugate mixed convection inside a concentrated photovoltaic thermal receiver. Compared to the conventional LBM, the NDLBM reveals the relationships of dimensionless governing parameters based on macroscopic, microscopic, and mesoscopic length scales.

3.5.2 Nondimensional Lattice Boltzmann Method

In this section, we will introduce a NDLBM which is suitable for both direct simulations and porous media simulations [76]. The porous medium model which directly connects macroscopic drag with microscopic drag coefficients by a geometry factor [74] is extended to the present NDLBM to avoid limitations of the valid range of the permeability and the Forcheimer coefficient [68, 81] used in earlier LBM porous medium models [57, 67].

The NDLBM discussed here is developed from that developed by Su et al. [75] for mixed convection by replacing the governing parameters Re_ℓ with $\sqrt{Ra_\ell/Pr_\ell}$, adding an extra body force term for the volume averaged drag between solid surfaces and fluid, and adding a supplementary term for the convection term of the momentum equation due to the definition of Darcy velocity. For the direct simulations, the two extra body force terms are equal to zero, and the bounce back boundary conditions are applied on every solid surface to represent the no-slip velocity on microscopic solid structures. The mesoscopic length scale must be less than the microscopic scale, i.e., $\ell < d < L$, so that both microscopic and macroscopic fluid flow and heat transfer fields can be solved. However, for porous media simulations, the mesoscopic length scale can be as large and $d < \ell < L$.

Similar to the NDLBM in [75], we denote f^* as the dimensionless density distribution function and g^* as the dimensionless temperature distribution function. Using the computational length scale $\ell = \Delta x$, the velocity scale $U = \sqrt{g\beta\Delta T L}$, the time scale $\Delta t = \frac{\ell}{U}$, the temperature scale ΔT, and the density scale ρ_{f0}, we obtain the following dimensionless equations for conservation of momentum and energy respectively,

$$f_k^*(\mathbf{x}^* + \mathbf{c}_k^*, t^* + 1) = f_k^*(\mathbf{x}^*, t^*) - \frac{1}{\tau_f^*}\left(f_k^*(\mathbf{x}^*, t^*) - f_k^{eq*}(\mathbf{x}^*, t^*)\right) + F_k^*, \quad (3.20)$$

$$g_k^*(\mathbf{x}^* + \mathbf{c}_k^*, t^* + 1) = g_k^*(\mathbf{x}^*, t^*) - \frac{1}{\tau_g^*}\left(g_k^*(\mathbf{x}^*, t^*) - g_k^{eq*}(\mathbf{x}^*, t^*)\right) + Q_k^*. \quad (3.21)$$

The respective dimensionless density, velocity, and temperature can be obtained as follows,

$$\rho^* = \sum_k f_k^*, \quad (3.22)$$

$$\mathbf{u}^* = \sum_k f_k^* \mathbf{c}_k^* / \sum_k f_k^*, \quad (3.23)$$

$$T^* = \sum_k g_k^*. \quad (3.24)$$

The local equilibrium distribution functions for fluid flow and heat transfer are given by,

$$f_k^{eq*} = w_k \zeta \left(\mathbf{c}_k^*, \mathbf{u}^* \right) \rho^*, \tag{3.25}$$

and

$$g_k^{eq*} = w_k \zeta \left(\mathbf{c}_k^*, \mathbf{u}^* \right) T^*, \tag{3.26}$$

respectively, where $\zeta \left(\mathbf{c}_k^*, \mathbf{u}^* \right)$ is obtained based on the Bhatnagar–Gross–Krook (BGK) model [77],

$$\zeta \left(\mathbf{c}_k^*, \mathbf{u}^* \right) = \left[1 + \left(\frac{\mathbf{c}_k^* \cdot \mathbf{u}^*}{c_s^{*2}} \right) + \frac{1}{2} \left(\frac{\mathbf{c}_k^* \cdot \mathbf{u}^*}{c_s^{*2}} \right)^2 - \frac{1}{2} \left(\frac{\mathbf{u}^* \cdot \mathbf{u}^*}{c_s^{*2}} \right) \right]. \tag{3.27}$$

For the two-dimensional nine discrete velocity direction (D2Q9) lattice mesh, the NDLBM speed of sound is $c_s^* = 1/\sqrt{3}$, according to $c_s^2 = c^2/3$ [28]. The corresponding dimensionless discrete velocities are

$$\mathbf{c}_k^* = \begin{cases} (0,0), & k = 0, \\ (\pm 1, 0), & k = 1, 2, 3, 4, \\ (\pm 1, \pm 1), & k = 5, 6, 7, 8, \end{cases} \tag{3.28}$$

and the weighting factors are,

$$w_k = \begin{cases} 4/9, & k = 0 \\ 1/9, & k = 1, 2, 3, 4, \\ 1/36, & k = 5, 6, 7, 8. \end{cases} \tag{3.29}$$

The discrete velocities and the weighting factors in the corresponding directions are shown in Fig. 3.2.

Based on the Chapman–Enskog expansion of the dimensionless Navier–Stokes and energy equations, similar to the analysis with units [77], we obtain the dimensionless relaxation times for the flow τ_f^* and heat transfer τ_g^* based on Re_ℓ and $Pe_\ell = Re_\ell Pr_\ell$ as discussed in [75]. In natural convection, the computational velocity scale is the natural convection velocity scale, hence $Ri_{ref} = 1$, and $Ra_\ell = Ri_{ref} Pr_\ell Re_\ell^2$; $Re_\ell = (Ra_\ell/Pr_\ell)^{1/2}$. The dimensionless relaxation times expressed in the form of the mesoscopic Rayleigh number Ra_ℓ and mesoscopic Prandtl number Pr_ℓ are,

$$\tau_f^* = \frac{1}{c_s^{*2}(Ra_\ell/Pr_\ell)^{1/2}} + \frac{1}{2}, \tag{3.30}$$

and

$$\tau_g^* = \frac{1}{c_s^{*2}(Ra_\ell Pr_\ell)^{1/2}} + \frac{1}{2}; . \tag{3.31}$$

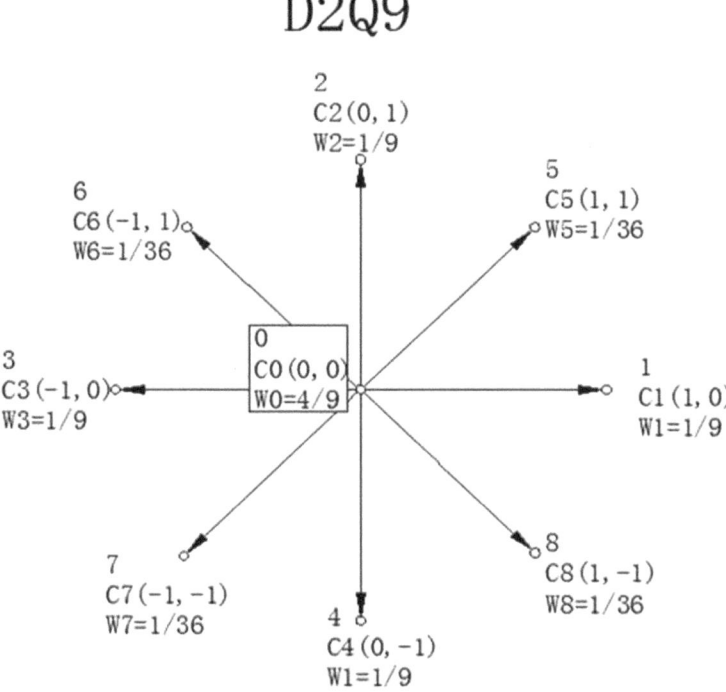

Fig. 3.2 Sketch of discrete velocities and the weighting factors for D2Q9 scheme

In order to keep the dimensionless relaxation times on the same order, a higher macroscopic Rayleigh number Ra_L cases need a finer mesh size ℓ for stable convergence.

Referring to the Navier–Stokes form of macroscopic governing equations, we can see that for porous medium models solved by the NDLBM, the dimensionless Darcy velocity $\mathbf{v}^* = \phi \hat{\mathbf{u}}^*$ should be used in Eqs. (3.23)–(3.27) to replace the microscopic fluid velocity $\mathbf{u}^*$. The dimensionless Darcy velocity $\mathbf{v}^*$ directly reveals the macroscopic flow, making it easer for use than previous LBM-based porous media models [21, 67]. To guarantee the conservation of momentum for the porous media model with the NDLBM,

$$M_k^* = -\mathbf{c}_k^* \cdot \left(\mathbf{v}^* \cdot \nabla \mathbf{v}^*\right), \tag{3.32}$$

because the convection term $\frac{1}{\phi}\mathbf{v}^* \cdot \nabla \mathbf{v}^*$ of the momentum equation has an extra $\frac{1}{\phi}$, compared to the momentum equation for pure fluid [74]. Previous LBM porous medium models have ignored the supplementary term of the convection term $-\mathbf{c}_k^* \cdot (\mathbf{v}^* \cdot \nabla \mathbf{v}^*)$ [21, 67], which is a valid assumption only as $\phi \to 1$.

The dimensionless force term due to the buoyancy force and the drag between the solid and fluid interfaces of the porous medium are expressed,

$$
F_k^* = \begin{cases} w_k \phi(\frac{-\mathbf{c}_k^* \cdot \mathbf{e}_g}{c_s^{*2}})T^*, & \text{for } \phi = 0 \text{ or } 1, \\ w_k \phi(\frac{-\mathbf{c}_k^* \cdot \mathbf{e}_g}{c_s^{*2}})T^* + w_k \frac{(1-\phi)}{\phi}(B_k^* + M_k^*), & \text{for } 0 < \phi < 1. \end{cases} \tag{3.33}
$$

where $\mathbf{e}_g$ is the unit vector in the gravity direction. For simulations, the dimensionless force term F_k^* will include a body force term due to the drag in form of $w_k \frac{(1-\phi)}{\phi} B_k^*$ and an extra supplementary term for convection term $w_k \frac{(1-\phi)}{\phi} M_k^*$. Based on the volume averaged drag term of [74], B_k^* of the present NDLBM is,

$$
B_k^* = -\frac{\eta}{2}\left(\frac{L}{d}\right)(\mathbf{c}_k^* \cdot \mathbf{v}^*)\left[\frac{1}{\phi}c_{d_0}|\mathbf{v}^*| + \left(\frac{Pr_f}{Ra_L}\right)^{\frac{1}{2}}\left(\frac{L}{d}\right)c_{d_1} \right.
$$
$$
\left. + \left(\frac{Pr_f}{Ra_L}\right)^{\frac{1}{4}}\left(\frac{1}{\phi}\frac{L}{d}\right)^{\frac{1}{2}}c_{d_2}|\mathbf{v}^*|^{\frac{1}{2}}\right], \tag{3.34}
$$

where c_{d_0}, c_{d_1}, and c_{d_2} are coefficients of the microscopic drag discussed in Chap. 2. The previous LBM models based on permeability K and the Forcheimer coefficient C_F [21, 67] have some limitations of range of porosity and geometries. Only for porous media materials with porosities greater than 0.9, the effects of porosity on K and C_F can be neglected [81]. It is also pointed out that the porosity effect on K and C_F should be considered for a low porosity porous materials [81]. The present NDLBM method is based on the model of [74], which is also discussed in Chap. 2. The permeability and the Forcheimer coefficients are related to porosity and the geometry factor as $K = \frac{\phi^2}{(1-\phi)\eta}\frac{2d^2}{c_{d_1}}$, and $C_F = \frac{\sqrt{(1-\phi)\eta}}{\phi^2}\frac{c_{d_0}}{\sqrt{2c_{d_1}}}$. The present closure model for the macroscopic drag is directly related to microscopic drag coefficients, and is thus applicable to wider ranges of the governing parameters.

The dimensionless heat source term is,

$$
Q_k^* = (1 - \phi)w_k \frac{q}{\rho c_p \Delta T U / \ell}, \tag{3.35}
$$

where q is the heat source in the solid.

Chapter 4
Illustration of Numerical Approaches

Abstract In this chapter, a 240-tube bundle heat exchangers immersed in a titled thin enclosure is simulated to show a practical application of the models and numerical methods discussed in Chaps. 1–3. The governing parameters in the macroscopic, microscopic, and mesoscopic length scales corresponding to the enclosure width, tube diameter, and mesh size are obtained. Porous medium model simulations by projection method and nondimensional lattice Boltzmann method (NDLBM), and direct simulations by NDLBM are compared for the transient energy discharged, the Nusselt numbers, the distributions of isotherms and streamlines, and the CPU times. Given the same grid number and simulation time, the CPU time of the porous medium model simulations by using NDLBM is about 1/60 of that of porous medium simulations by using finite difference based on the projection method, and 1/20 of that of the direct simulations with the uniform code based on NDLBM. The porous medium simulations can only obtain Darcy velocity and volume averaged temperature, while the direct simulations can obtain both macroscopic and microscopic velocity and temperature.

Keywords Mesoscopic length scale · Nondimensional lattice Boltzmann method · Projection method

4.1 Problem Definition

Transient fluid flow and heat transfer in tube bundle heat exchangers immersed in water-filled enclosures have application in solar water storage tanks [45, 70]. Early work to understand convective heat transfer focused on numerical modeling [44]. This unit cell approximation for heat exchangers usually combines periodic boundary conditions that may not consider the effects of random distributions of pore networks [9]. It is time- and space-consuming to do direct simulations on the transient flow and heat transfer on bundle of hundreds of small diameter tubes by conventional computational methods such as finite difference, finite element, and control volume methods. The computational speed and accuracy of these methods are highly dependent on convergence speed and accuracy of the Poisson solver [47]. The computational time is dramatically increased for high grid numbers, which are required to capture the physicals in practical tank size.

© The Authors 2015   27
Y. Su, J. H. Davidson, *Modeling Approaches to Natural Convection in Porous Media*,
SpringerBriefs in Applied Sciences and Technology, DOI 10.1007/978-3-319-14237-1_4

Treatment of a tube bundle as a porous medium based on an represented elementary volume (REV) is presented and applied in [69, 74]. However, only the macroscopic flow and heat transfer, i.e., the Darcy velocity and volume averaged temperature in the REV, could be obtained. The results did not cover the effects of different distributions of tubes with the same porosity and geometry factors. Recently, both porous medium model and direct simulations by lattice Boltzmann method (LBM) have been developed. Also, the nondimensional lattice Boltzmann method (NDLBM) developed by Su et al. [75] makes it possible for the comparison of the numerical results and the computational speeds for three numerical approaches discussed in Chap. 3. In this chapter, we first introduce the geometries of the illustrative example, i.e., a 240-tube bundle heat exchanger immersed in a thin titled enclosure. Then, the details of the three numerical approaches for solving the example are presented in Sect. 4.2. In Sect. 4.3, we compare the results obtained by the three approaches including the transient averaged temperature, the energy discharged, the transient Nusselt numbers, and the transient isotherms and streamlines. Finally, the CPU times are compared to show the computational speeds of the three numerical approaches.

The computational domain is based on a laboratory enclosure developed by Liu et al. [46] for evaluation of an integral collector storage system with an immersed heat exchanger. The aspect ratio is 9:1, and the enclosure is inclined at 30° to the horizontal. A 240-tube bundle is positioned near the top of the tank. Each tube has an outer diameter of 3.2 mm, inside diameter of 1.8 mm, and a length of 1 m. The tubes are arranged in parallel with $P/D = 3.3$. They are staggered with respect to the primary convective flow pattern. The heat exchange surface is 2.38 m^2 including the headers. The heat exchanger is used to extract energy from the collector by flowing with inlet temperature ($T_{in} = 60°$). The model treats the storage thin enclosure as a multiple-zone enclosure. As shown in Fig. 4.1, the tank is of height $k_1 L$, where L is the width of the tank. In this study, $L = 0.1$ m. The ratio between the width of the tank and the tube diameter is $L/d = 31.25$. The heat exchanger is treated as a porous medium zone of height $k_2 L$. The heat exchanger is positioned at a distance $k_3 L$ from the top of the tank. The geometry factors are $(k_1, k_2, k_3) = (9, 3, 1)$.

4.2   Application of the Numerical Methods

For the present problem, the macroscopic length scale is the width of the tank L and the microscopic length scale is the tube diameter d. The enclosure domain is meshed in size of $m \times n$, then the lattice mesh size, the length scale based on the mesh size is $\ell = \Delta x = \Delta y = L/m = H/n$. The temperature is scaled by the initial heat flux q_0 as $\Delta T = \left[\sqrt{\frac{L}{g\beta} \frac{q_0}{(\rho c_p)_f}} \right]^{\frac{2}{3}}$, and the reference temperature is the averaged initial temperature T_0.

Fig. 4.1 Multizone enclosure
model of a thermal storage
thin enclosure with an
immersed heat exchanger. **a**
Rectangular storage thin
enclosure with an immersed
tube bundle for discharge. **b**
Multizone porous media
model in which the heat
exchanger is treated as a
porous medium

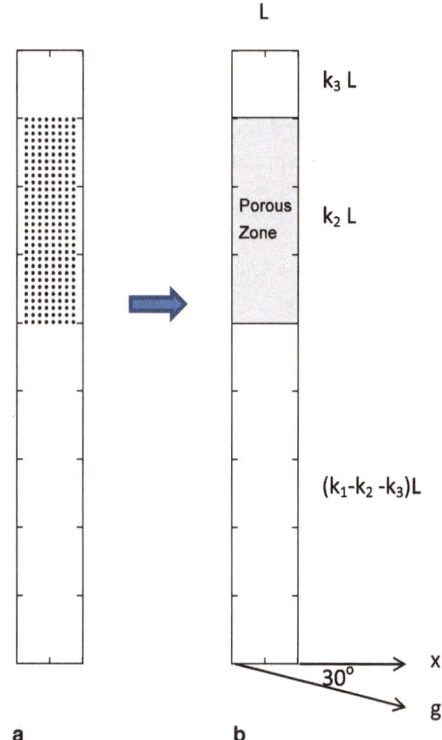

4.2.1 Porous Medium Approach by Projection Method

The bulk porosity in the region occupied by the heat exchanger is,

$$\phi = 1 - \frac{N\pi(d/2)^2}{k_2 L^2} \quad or \quad \phi = 1 - \pi\left(\frac{d}{2P}\right)^2. \tag{4.1}$$

In this simulated geometry, $\phi = 0.9279$. The region with no heat exchanger, i.e., the pure fluid region, is treated as a nonporous zone, i.e., $\phi = 1$. The local thermal equilibrium (LTE) heat transfer model discussed in Chap. 1 is used to model the transient discharge process [46]. The heat exchanger tubes are modeled as a volumetric heat source based on the energy balance,

$$\dot{m}C_p(T_{in} - T_{out}) = q V_{tubes}. \tag{4.2}$$

In actual operation, the heat flux varies in the flow direction and with time as energy is discharged (or added) from the storage tank. Prior measurements of overall heat transfer [46] suggest that spatial and temporal variations can be decoupled, and thus q can be expressed as the product,

$$q = q_0 F(z)G(t), \tag{4.3}$$

where q_0 is an initial value. In general, q_0 can be estimated for specified operating conditions, initial tank temperature, and heat exchanger geometry. In the present model, $q_0 = -4.9341 \times 10^6$ W/m^3. The spatial variation of q_0 from the inlet to the outlet of the tubes is given by $F(z)$. The temporal variation is given by $G(t)$. For a three-dimensional model based on the exponential decay assumption, the spatial function can be expressed as

$$F(z) = \frac{k_3 exp\,(-z/L)}{1 - exp(-k_3)}. \tag{4.4}$$

In two dimensions,

$$F(z) = 1. \tag{4.5}$$

In two dimensions, the dimensionless heat sink term used to model the heat exchanger is,

$$q^* = \frac{qL}{(\rho C_p)_f U \Delta T} = \frac{q}{|q_0|} = G^*(\tau). \tag{4.6}$$

In the derivation of $G^*(\tau)$, it is useful to define the average Nusselt number of the heat exchanger using the average temperature of the fluid in the tank, $\widehat{\overline{T}}_f$, as the reference temperature,

$$Nu_d = \frac{hd}{k_f} = \frac{qd/4}{\widehat{\overline{T}}_s - \widehat{\overline{T}}_f} \frac{d}{k_f}. \tag{4.7}$$

Using an energy balance on the enclosure,

$$\rho_f C_p V_T \frac{d\widehat{\overline{T}}_f}{dt} = \frac{d}{4} q N A_o, \tag{4.8}$$

and the Nusselt number can be expressed as,

$$Nu_d = \frac{(1 - \phi)\frac{V_{porous}}{V_{tank}} \eta \left(\frac{Ra_d}{Ra_L}\right)^{1/6} Pr_f^{1/2} Ra_d^{1/2}}{\widehat{\Theta}_s - \widehat{\Theta}_f} \frac{\partial \overline{\widehat{\Theta}}_f}{\partial \tau}, \tag{4.9}$$

where the average dimensionless temperature of the storage volume is $\overline{\widehat{\Theta}}_f = \frac{\widehat{\overline{T}}_f - T_0}{\Delta T}$. The dimensionless temperature in the porous (heat exchanger) zone is $\widehat{\Theta}_s = \frac{\widehat{\overline{T}}_s - T_0}{\Delta T}$. Based on scale analysis

$$\frac{d\overline{\widehat{\Theta}}_f}{Nu_d(\overline{\widehat{\Theta}}_f - \widehat{\Theta}_s)} \sim \frac{dq}{q} = \frac{dG^*(\tau)}{G^*(\tau)}. \tag{4.10}$$

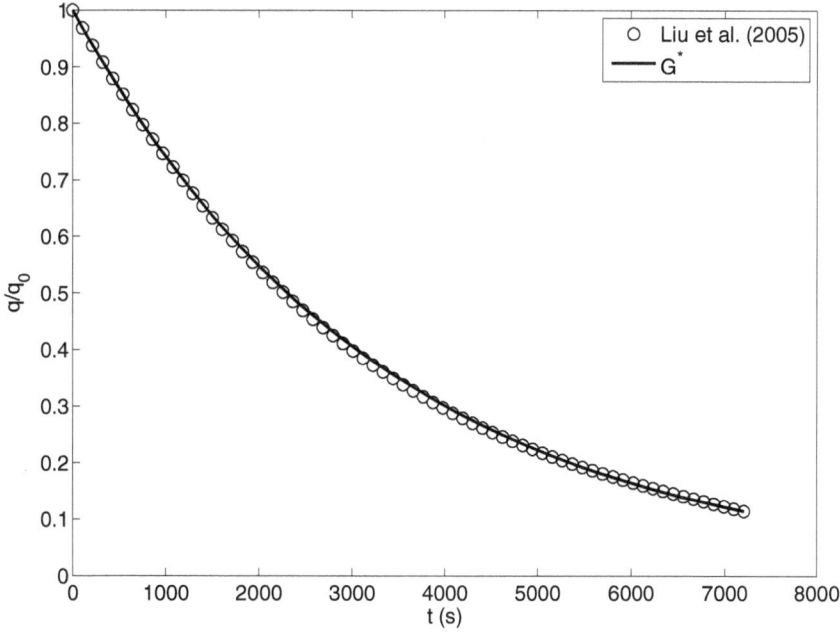

Fig. 4.2 Correlation for constant $C = 0.26$ in G^*

Substitution of Eq. (4.9) into (4.10), rearranging, and then integrating from 0 to τ yields an expression for $G^*(\tau)$,

$$G^*(\tau) = \frac{q_0}{|q_0|} exp \left(\frac{-\tau}{(1-\phi)\frac{V_{porous}}{V_{tank}} \eta(\frac{d}{L})^{1/2} Pr_f^{1/2} Ra_d^{1/2}}{C} \right). \qquad (4.11)$$

As shown in Fig. 4.2, for the rectangular tank with a 240-tube heat exchanger, $C = 0.26 \pm 0.01$ based on a curve fit of prior measurements of overall heat transfer rate [46].

The governing parameters are $(Ra_L, Ra_d, Pr_f, \phi_p) = (1.2 \times 10^8, 4.0 \times 10^3, 7.0, 0.9279)$. Initially, the storage fluid is assumed to be quiescent and fully mixed (isothermal), and therefore,

$$\mathbf{v}^* = 0 \ and \ \hat{\Theta} = 0 \quad at \ \tau = 0. \qquad (4.12)$$

Boundary conditions at the tank wall, which is assumed to be adiabatic with no slip, are,

$$\mathbf{v}^* = 0 \ and \ \frac{\partial \hat{\Theta}}{\partial \mathbf{n}} = 0. \qquad (4.13)$$

Solution yields transient values of the temperature and velocity distributions, as well as average Nusselt number.

4.2.1.1   Code Validation for Randomly Packed Spheres

As a standard for the comparison of the heat transfer for various cases, the Nusselt number on top of the enclosure is based on the macroscopic length scale H, i.e., the height of the enclosure, and is defined as,

$$Nu_H = \frac{\int_0^1 2\pi r^* \frac{\partial \hat{T}^*}{\partial z^*} d\, r^*}{\int_0^1 2\pi r^* d\, r^*}, \quad at \quad z^* = 1, \tag{4.14}$$

where (r^*, z^*) are cylindrical coordinates. The porous medium Nusslet number is defined as,

$$Nu_m = \frac{k_f}{k_m} Nu_H. \tag{4.15}$$

We find that at low values of Rayleigh number, steady Nusselt numbers are observed, while at large Rayleigh numbers, fluctuations occur, and a periodic behavior is seen in flow patterns and overall heat transfer coefficients at the upper surface. Thus, a time averaged Nusselt number based on the characteristic period of oscillation with dimensionless period t_0^* is given by,

$$\overline{Nu_H} = \frac{\int_{t^* - t_0^*/2}^{t^* + t_0^*/2} Nu_H \, dt^*}{t_0^*}. \tag{4.16}$$

Similarly, the time averaged porous medium Nusselt number is denoted by $\overline{Nu_m}$.

First, validation of our solution method was obtained by solving the Rayleigh–Bénard problem for a vertical cylinder filled with water ($Pr_f \sim 7$). Results in terms of average Nusselt number at the upper surface were obtained for $8.9 \times 10^5 < Ra_H < 2.6 \times 10^8$. The present results were compared favorably with three correlations [37, 85, 22] as shown in Fig. 4.3. Also Fig. 4.3 shows that $log(\overline{Nu_H})$ is not linear in $log(Ra_H)$ in the range $2.5 \times 10^3 < Ra_H < 10^5$. The onset of convection for the water-filled cylindrical enclosure is $Ra_H \sim 2.5 \times 10^3$, larger than 1708 for the infinite layer via linear stability theory owing to the finite aspect ratio of the solution domain [23]. This slight delay is due to the enhancement of the effective viscosity by the cylindrical wall [72] and the delay of onset Rayleigh number is also observed by enhanced effective thermal conductivity and enhanced effective viscosity by [43].

Validation of our solution method was then obtained by solving the Rayleigh–Bénard problem for a vertical cylinder filled with spheres. The spheres are filled either in full or in the bottom half of the enclosure. The diameter of the cylinder is D, the height of the enclosure is H, H_p is the height of the porous layer, and H_p/H is the ratio of the height of the porous layer over the total height of the enclosure. Nusselt numbers for the cylindrical solution domain fully packed with spheres (random stacking) are compared to those of the prior studies that were using the Darcy Brinkman formulation with Ergun constant $a = 175$ and $b = 1.75$ for $10^3 < Ra_H < 10^{11}$, $D/H = 2$, and $Da = 10^{-4}$. Figure 4.4 shows that at $Ra_m < 10^3$

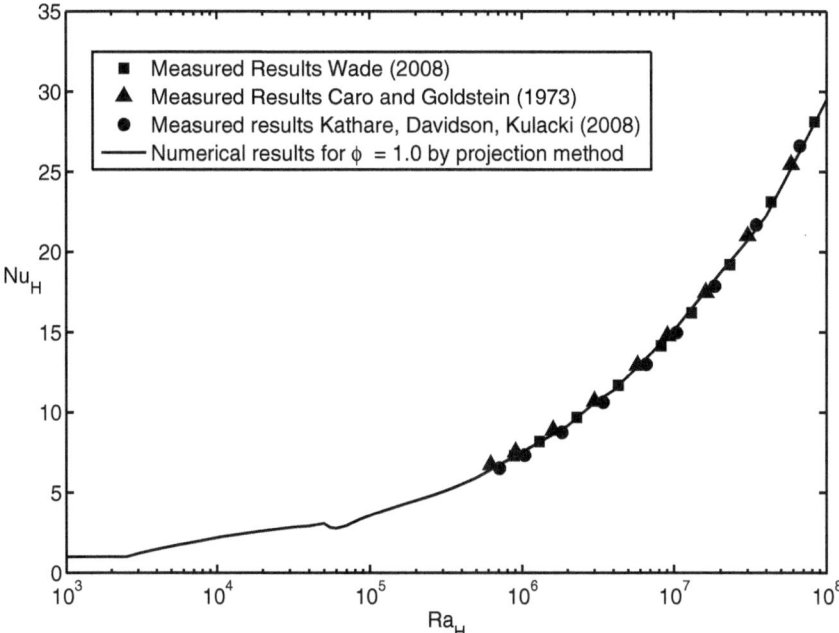

Fig. 4.3 Nusselt number for Rayleigh–Bénard convection in a cylindrical enclosure

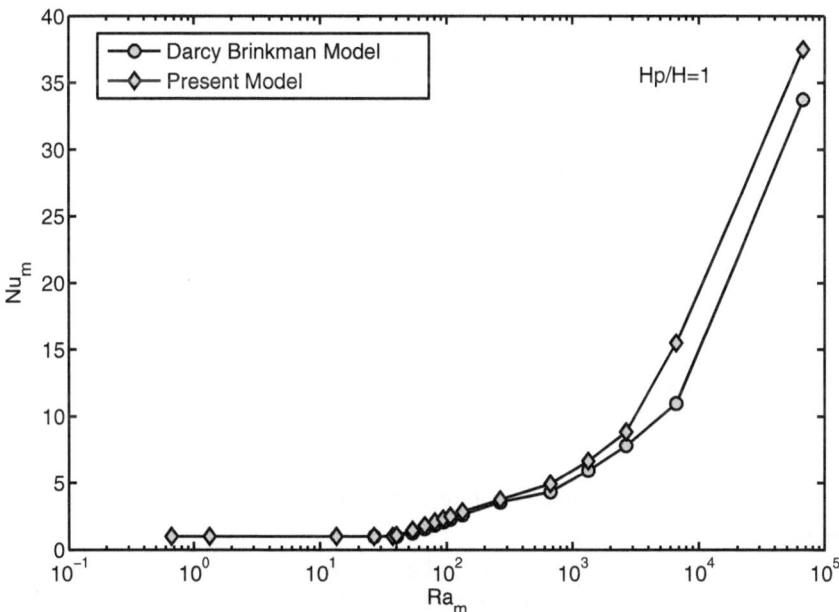

Fig. 4.4 Nusselt numbers for a bed of packed spheres via the present drag model and the Darcy–Brinkman–Forcheimer formulation

and $Ra_m > 5 \times 10^6$, the two models yield almost identical results, while at middle range of values of Ra_m the present model predicts slightly higher value of $\overline{Nu}_m$. The higher values are attributed to the third term of the drag model. As shown in Fig. 4.4, the shape of the curve agrees with the study of Bagchi and Kulacki [3]. There are some sudden changes of slope in the curves and this change of slope at lower Ra_H is also observed by Poulikakos [61]. Also shown in Fig. 4.4, for $H_p/H = 1$, the onset of convection occurs at $Ra_m \approx 40$, and for $H_p/H = 0.5$, the onset point is slightly delayed.

4.2.2 Porous Medium Approach by NDLBM

For the porous medium approach by the NDLBM, the sketch of the porous medium model is the same as Fig. 4.1b. The porosity is also defined as,

$$\phi = \begin{cases} 1, & \text{for fluid part,} \\ \phi_p = 1 - \frac{N\pi(d/2)^2 L_z}{V_{porous}}, & \text{for porous medium part,} \end{cases} \tag{4.17}$$

where N is the tube number, V_{porous} is the porous zone volume, L_z is length in the third direction, and d is the tube diameter, i.e., the microscopic length scale.

As discussed in Chap. 3, we denote f^* as the dimensionless density distribution function and g^* as the dimensionless temperature distribution function. Using the computational length scale $\ell = \Delta x$, the velocity scale $U = \sqrt{g\beta\Delta T L}$, the time scale $\Delta t = \frac{\ell}{U}$, the temperature scale $\Delta T = \left[\sqrt{\frac{L}{g\beta}}\frac{q_0}{(\rho c_p)_f}\right]^{\frac{2}{3}}$, and the density scale ρ_f, we can obtain the following dimensionless equations for the momentum equation and heat transfer equation respectively.

The D2Q9 scheme is chosen in the present simulation because direct simulations of the 240 tubes by the NDLBM is still time consuming, although it is much quicker than conventional finite difference and finite volume method. Also previous studies show that two-dimensional simulations can capture the heat transfer coefficient almost equal to those of three-dimensional study, although for fluid patterns, two-dimensional simulations can only show two-dimensional recirculations rather than three-dimensional spirals [73]. For the present problem, the macroscopic length scale is the width of the tank L. The inside tank domain is meshed in size of $m \times n$, then the lattice mesh size, the length scale for NDLBM simulations, is $\Delta x = \Delta y = L/m = H/n$. The temperature scale is scaled by the initial heat flux q_0 as $\Delta T = \left[\sqrt{\frac{L}{g\beta}}\frac{q_0}{(\rho c_p)_f}\right]^{\frac{2}{3}}$, and the reference temperature is the averaged tank initial temperature T_0.

Based on scale analysis of the overall energy balance equation in the enclosure [69, 70], the dimensionless heat source term can be,

$$Q_k^* = (1-\phi)w_k\frac{q_0}{|q_0|}\exp\left[\frac{-C\eta(1-\phi)(\frac{V_{porous}}{V_{tank}})t^*}{\left(\frac{dL}{\ell^2}Pr_f Ra_d\right)^{1/2}}\right] \tag{4.18}$$

In the present study, the shape of the heat exchangers are cylinders, and the geometry factor $\eta = 4$ [74]. Based on a curve fit of prior measurements of overall heat transfer during discharge [45, 69], $C = 0.26 \pm 0.01$.

In the porous medium approach by the NDLBM, there is no need to apply the bounce back conditions for each of the 240 tubes.

Similar to the no-slip boundary conditions of [32], bounce back boundary conditions are applied on the storage walls as no slip zero velocity conditions,

$$f_k^* = f_{k_{od}}^*, \quad k = 0 \sim 8, \tag{4.19}$$

where k_{od} is the opposite direction of k. The bounce back conditions are also applied for each of the 240 tubes for direct numerical simulations.

For dimensionless temperature boundary conditions, adiabatic temperature boundary conditions are applied to the walls as,

$$g_{k,wall}^* = g_{k,wall-1}^*. \tag{4.20}$$

At the initial time, the tank is assumed to be at a uniform initial temperature, which is also the reference temperature. Zero initial velocity and uniform environment temperature are selected as the initial conditions for flow and heat transfer fields. The corresponding equilibrium distribution functions are,

$$f_k^* = w_k \rho_0^*, \quad \text{at } t^* = 0, \tag{4.21}$$

and

$$g_k^* = w_k T_0^*, \quad \text{at } t^* = 0. \tag{4.22}$$

Thus, $\rho_0^* = \phi + (1 - \phi)(\rho_s/\rho_f)$ and $T_0^* = 0$ are applied for the present study.

4.2.3 Direct Simulation Approach by NDLBM

For the direct simulation approach by the NDLBM, the sketch of the porous medium model is the original one shown in Fig. 4.1a.

For direct simulation model by the NDLBM,

$$\phi = \begin{cases} 1, & \text{for the fluid,} \\ 0, & \text{for the solid,} \end{cases} \tag{4.23}$$

The heat source is in the solid and in similar form as Eq. (4.18),

$$Q_k^* = (1 - \phi)w_k \frac{q_0}{|q_0|} \exp\left[\frac{-C\eta(1 - \phi_p)(\frac{V_{porous}}{V_{tank}})t^*}{\left(\frac{dL}{\ell^2} Pr_f Ra_d\right)^{1/2}}\right]. \tag{4.24}$$

Noting that the ϕ_p in Eq. (4.17) is used in Eq. (4.24) to maintain the same constant C applied in both porous medium and direct simulations. The bounce back conditions are also applied for each of the 240 tubes for direct numerical simulations.

A computational code suited to both direct simulation and porous medium model simulation based on the present NDLBM is developed in Fortran 90, which is compiled and run on the high performance computing cluster with GNU/Linux operating system (CentOS 5.5 64-bit). To maintain high accuracy, the mesh size $n \times m$ is set to be 200×1800, and the lattice size Δx is smaller than the required size for the finite difference based projection method [74, 69]. The models shown in Fig. 4.1 are computed based on the same governing parameters: macroscopic Rayleigh number $Ra_L = 1.24 \times 10^8$ (or $Ra_H = 9.1 \times 10^{10}$), microscopic Rayleigh number $Ra_d = 4.1 \times 10^3$, mesoscopic Rayleigh number $Ra_\ell = 15.54$, and Prandtl number of the fluid $Pr_f = 7.0$. The solid fluid thermal conductivity and heat capacity ratio are $k_s/k_f = 668.3$ and $(\rho c_p)_s/(\rho c_p)_f = 0.8213$, respectively.

4.3 Comparison of Results

4.3.1 Tank Averaged Temperature and Energy Discharged

In order to validate the energy conservation of the present code, the predicted transient tank averaged temperature and the fraction of energy discharged are compared to experimental and numerical studies in Fig. 4.5. The transient dimensionless temperature (Fig. 4.5a) drops dramatically for the first 1000 s and then decreases more slowly. The energy is removed from the tank.

Both the transient tank averaged temperature and fraction of energy discharged of the direction simulation, and the porous medium model simulation by NDLBM agree with those obtained by the experimental study of [45] and the porous medium simulation by finite difference-based projection method of Su et al. [70]. This result confirms energy conservation of the uniform computational code, which validates the present NDLBM for both direct simulations and porous medium model simulations.

4.3.2 Transient Averaged Nusselt Numbers of Tubes

Referring to the scale analysis of [70], the transient averaged Nusselt number for the 240 tubes is estimated by the transient averaged temperatures in different zones as,

$$\overline{Nu_d} = \frac{\eta(1 - \phi_p)(\frac{V_{porous}}{V_{tank}})\left(\frac{dL}{\ell^2}Pr_f Ra_d\right)^{1/2}}{(T^*_{porous} - T^*_{tank})}\frac{\partial T^*_{tank}}{\partial t^*}. \tag{4.25}$$

Equation (4.25) shows that a better designed heat exchanger will have a smaller temperature difference $T^*_{porous} - T^*_{tank}$, i.e., the temperature will be more evenly distributed in both heat exchanger and pure water zones. The transient averaged Nusselt

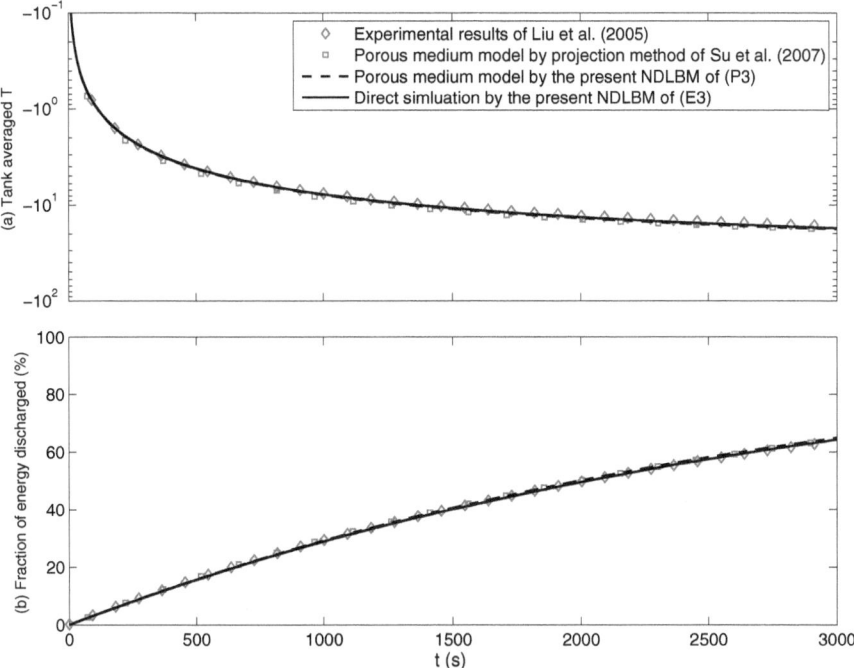

Fig. 4.5 The transient dimensionless tank averaged temperature and ratio of energy discharged

number, $\overline{Nu_d}$, for the 240 tubes can be calculated from the transient temperature results in Eq. (4.25). Transient averaged Nusselt number, $\overline{Nu_d}$, for the 240 tubes obtained from the direct simulation by NDLBM and the porous medium model simulation by NDLBM, and the porous medium model simulation by projection method [69] are compared with experimental study of [46] in Fig. 4.6. As shown in this figure, the transient averaged Nusselt number $\overline{Nu_d}$ decays with time, and approaches a stable value near 10s consistent with the experimental measurement of [46]. Also shown in Fig. 4.6, the direct simulation obtained by the NDLBM model of Fig. 4.1a is more stable than the results obtained by the porous medium model of Fig. (4.1b). However, the simulation speed of porous medium model is about ten times faster than direct simulations by the uniform NDLBM code with the same grid numbers.

4.3.3 Transient Isotherms and Streamlines

Figure 4.7 compares the transient dimensionless streamlines and isotherms when the first macroscopic thermal plume is generated for the three approaches. From this figure, we can see that the first macroscopic mushroom structure thermal plume drops along the back wall of the thin enclosure. Both the direct simulation by NDLBM case

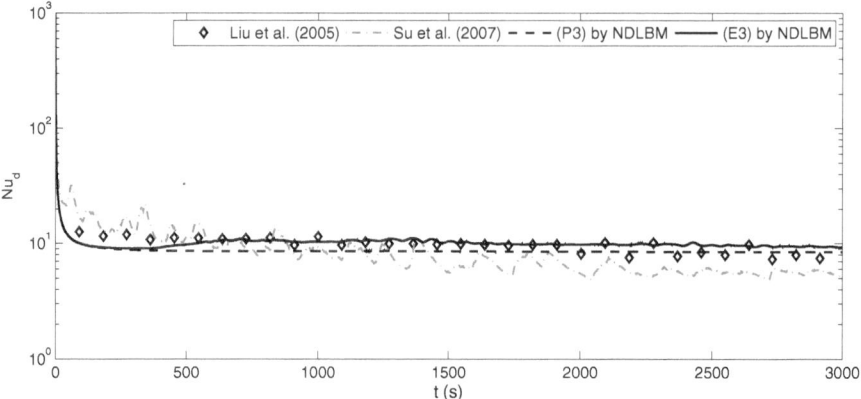

Fig. 4.6 Comparison of the transient averaged Nusselt number, $\overline{Nu_d}$, for the 240 tubes to previous studies

and the porous medium model simulations by NDLBM and projection approach show the macroscopic mushroom structure thermal plume. Thus, both direct simulations and porous medium model simulations with the NDLBM can obtain the macroscopic fluid flow and heat transfer. However, the porous medium model simulation result shown in Fig. 4.7 does not capture the microscopic flows and temperature distributions between the tubes, due to the fact that the governing equations are based on volume-averaged values in a REV.

Next, we compare the isotherms during the discharge process for the three approaches. We can see from Fig. 4.8 that the macroscopic flow pattern and the temperature distributions are similar. Only the direct simulation by NDLBM shows the microscopic flow and heat transfer. Both porous medium model simulations and the direct numerical simulations by NDLBM show that there is a stratification below the heat exchanger. The temperature stratification is not very clear for the porous media model simulation by projection method.

4.3.4   Comparison of CPU Times

The CPU time for the models are compared in Table 4.1. From top to bottom, the models are the porous medium simulations with finite difference-based projection method [69], porous medium simulations with NDLBM, and the direct simulation with NDLBM. The CPU time for the porous medium simulations by using the NDLBM is 1/60 the CPU time required by the finite difference-based projection method. The finite difference-based projection method must complete many time-consuming iterations to obtain the solutions for Poisson equations of the pressure field

Fig. 4.7 Comparison of
dimensionless isotherms and
streamlines for the three
approaches

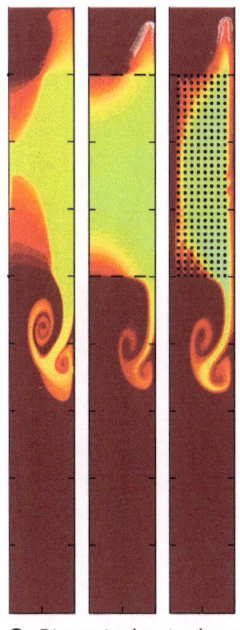

a Dimensionless isotherms

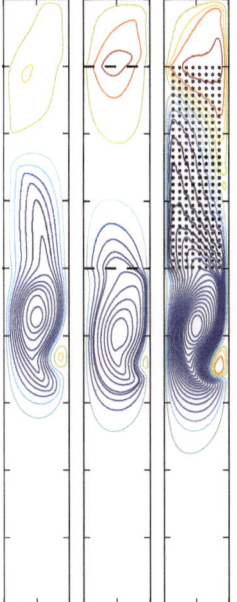

b Dimensionless streamlines

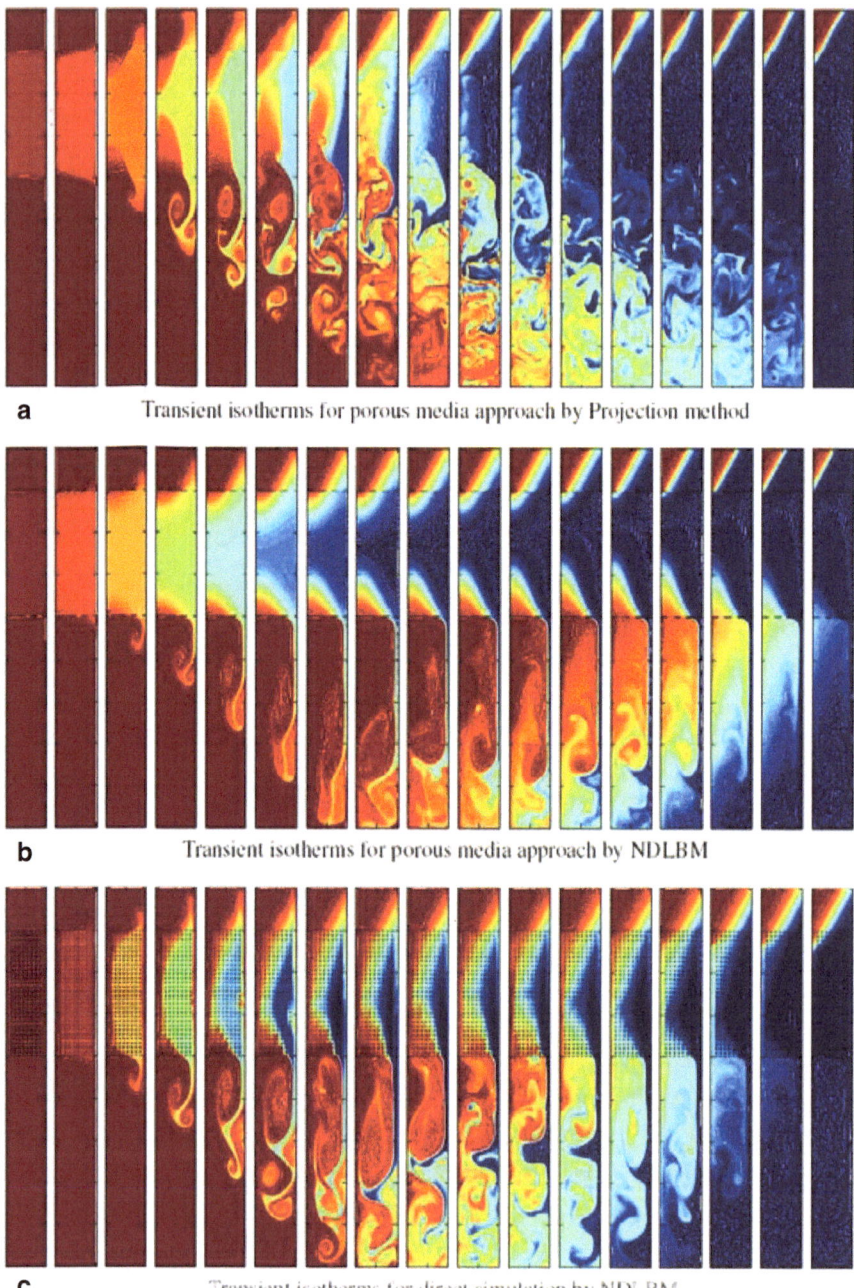

a   Transient isotherms for porous media approach by Projection method

b   Transient isotherms for porous media approach by NDLBM

c   Transient isotherms for direct simulation by NDLBM

Fig. 4.8 Comparison of transient isotherms of the three kinds of simulations

Table 4.1 Comparison of CPU time

Model	Method	Grid number	dt		Simulation time	CPU time	CPU time
			(s)		(s)	(dd:hh:mm:ss)	(s)
Fig. 4.1b	FDM [74]	200×1800	1.1866×10^{-3}		1.1866×10^2	04:00:04:53	345,893
Fig. 4.1b	FDM [74]	200×1800	1.1866×10^{-3}		1.1866×10^3	52:02:29:10	4,501,750
Fig. 4.1b	NDLBM	200×1800	1.1866×10^{-2}		1.1866×10^3	00:20:11:30	72,690
Fig. 4.1b	NDLBM	400×3600	5.9331×10^{-3}		5.9331×10^2	03:08:59:15	291,555
Fig. 4.1a	NDLBM	200×1800	1.1866×10^{-2}		1.1866×10^3	17:22:32:50	1,549,970
Fig. 4.1a	NDLBM	400×3600	5.9331×10^{-3}		5.9331×10^2	69:01:37:00	5,967,420

at each time step, whereas the NDLBM only needs to solve linear equations. Moreover, note that the time step used in the finite difference-based projection method is often required to be small in order to guarantee the convergence of the Poisson solver. When the time step is relatively large, this method may not converge. Instead, the NDLBM can use a relatively large time step.

Moreover, it can be seen from Table 4.1 that the CPU time for porous medium simulations can be considerably smaller than that for the corresponding direct simulations. For example, the CPU time for porous media model by the NDLBM is 1/20 of the CPU time for direct simulation model by the same NDLBM. The larger number of boundary conditions required by the direct simulation are responsible for the increased CPU time. The CPU time by the NDLBM is linear with respect to the grid numbers for both direct simulations and porous medium simulations.

4.3.5 Summary

Comparison studies based on natural convection of 240 cylindrical tube bundle heat exchangers immersed in a thin enclosure were performed to illustrate the computational efficiency and accuracy of both direct simulations and porous medium model simulations achieved by using NDLBM. Transient flow and heat transfer distributions, and the averaged Nusselt numbers for modes were compared for the results obtained by three kinds of numerical simulations. The NDLBM simulations are based on mesoscopic length scale Rayleigh number. When the mesoscopic length scale is smaller than the microscopic length scale, the microscopic flow and heat transfer nearby each tube (i.e., the real local velocity and temperature) as well as the macroscopic temperature and velocity distributions in the far field can be obtained by direct simulation. When the mesoscopic length scale is between the microscopic and macroscopic length scales, only the macroscopic scale flow and heat transfer (i.e., the Darcy velocity and volume averaged temperature) can be obtained by porous medium model simulations.

Given the same grid number and simulation time, the CPU time of the porous medium model simulations using the NDLBM is about 1/60 of that using the conventional finite difference method, and is about 1/20 of the CPU time for the direct simulations by using the present NDLBM. However, the porous medium model simulations can only provide macroscopic flow and heat transfer, while the direct simulations can clearly show both macroscopic and microscopic transient flow and temperature distributions.

References

1. Amir T, Koroush P, Alireza S, Hadi Ghezel S (2010) Spectral method for solving differential equation of gas flow through a micro-nano porous media. J Comp Theo Nanosci 7:542–546
2. Amaziane B, Bergam A, Ossmani M El, Mahazli Z (2009) A posteriori estimators for vertex cenred finite volume discretization of a convection-diffusion-reaction equation arising in flow in porous media. Int J Numer Meth Fluids 59:259–284
3. Bagchi A, Kulacki FA (2011) Natural convection in horizontal fluid-superposed porous layers heated locally from below. Int J Heat Mass Transf 54:3672–3682
4. Bear J (1972) Dynamics of fluids in porous media. Dover, New York
5. Bhattacharya A, Mahajan RL (2006) Metal foam and finned metal foam heat sinks for electronics cooling in buoyancy-included convection. ASME J Elec Packag 128:259–266
6. Braga EJ, de Lemos MJS (2004) Turbulent natural convection in a porous square cavity computed with a macroscopic $\kappa - \varepsilon$ model. Int J Heat Mass Transf 47:5639–5650
7. Brinkman HC (1947) A calculation of viscous force exerted by a flowing fluid on a dense swarm of particles. Appl Sci Res A1:27–34
8. Buck R (2000) Massenstrom-Instabilitäten bei volumetrischen Receiver-Reaktoren, Ph.D. thesis, Universität Stuttgart, Germany
9. Chattopadhyay A, Pattamatta A (2014) Energy transport across submicron porous structures: a lattice Boltzmann study. Int J Heat Mass Transf 72:479–488
10. Chen CK, Hsiao SW (1998) Transport phenomena in enclosed porous cavities. In: Ingham DB, Pop I (eds) Transport phenomena in porous media. Elsevier, Oxford, pp. 31–56
11. Chen L, Kang QJ, Mua Y, He YL, Tao WQ (2014) A critical review of the pseudopotential multiphase lattice Boltzmann model: methods and applications. Int J Heat Mass Transf 76:210–236
12. Chorin AJ (1968) Numerical solution of the Navier-Stokes equations. Math Comp 22:745–762
13. Churchill SW, Bernstein M (1977) A correlating equation for forced convection from gases and liquids to a circular cylinder in cross flow. ASME J Heat Transf 99:300–306
14. Ehlers W (2002) Foundations of multiphasic and porous materials. In: Ehlers W, Bluhm J (eds) Porous media theory: experiments and numerical applications. Springer, Hamburg, pp. 3–86
15. Davidson JH, Kulacki FA, Savela D (2009) Natural convection in water-saturated reticulated vitreous carbon foam. Int J Heat Mass Transf 52:4479–4483
16. Ergun S (1952) Fluid flow through packed columns. Chem Eng Prog 48:89–94
17. Escobar RA, Ghai SS, Jhon MS, Amon CH (2006) Multi-length and time scale thermal transport using the lattice Boltzmann method with application to electronics cooling. Int J Heat Mass Transf 49:97–107
18. Ferziger JH, Peric M (1999) Solution of the Navier-Stokes equations. In: Ferziger JH, Peric M (eds) Computational methods for fluid dynamics. Springer, Hamburg, pp. 206–208
19. Feiereisen TJ, Klein SA, Duffie JA, Beckman WA (1982) Heat transfer from immersed coils, Paper 82 WA/SOL-18, American Society of Mechanical Engineers, New York

© The Authors 2015
Y. Su, J. H. Davidson, *Modeling Approaches to Natural Convection in Porous Media*,
SpringerBriefs in Applied Sciences and Technology, DOI 10.1007/978-3-319-14237-1

20. Fu X, Viskanta R, Gore JP (1998) Measurement and correlation of volumetric heat transfer coefficients of cellular ceramics. Exp Therm Fluid Sci 17:285–293

21. Gao DY, Chen ZQ, Chen LH (2014) A thermal lattice Boltzmann model for natural convection in porous media under local thermal non-equilibrium conditions. Int J Heat Mass Transf 70: 979–989

22. Garon AM, Goldstein RJ (1973) Velocity and heat transfer measurements in thermal convection. Phys Fluids 16:1818–1825

23. Gebhart B, Jaluria Y, Mahajan RL, Sammakia B (1988) Buoyancy-induced flows and transport. Hemisphere, Washington DC

24. Guermond JL, Minev P, Shen J (2006) An overview of projection methods for incompressible flows. Comput Methods Appl Mech Eng 195:6011–6045

25. Guo Y, Bennacer R, Shen S, Ameziani DE, Bouzidi M (2010) Simulation of mixed convection in slender rectangular cavity with lattice Boltzmann method. Int J Numer Methods Heat Fluid Flow 20:130–148

26. Haltiwanger J, Davidson JH (2009) Discharge of a thermal storage tank using an immersed heat exchanger with an annular baffle. Sol Energy 83:193–201

27. Hazen A (1893) Some physical properties of sand and gravels with special reference to their use in filtration. Massachusetts State Board of Health, Twenty-fourth Annual Report 541

28. He X, Luo LS (1997) Theory of the lattice Boltzmann: from the Boltzmann equation to the lattice Boltzmann equation. Phys Rev E 56:6811–6817

29. Hsu CT, Cheng P (1990) Thermal dispersion in a porous medium. Int J Heat Mass Transf 33:1587–1597

30. Hsu CT (1999) A closure model for transient heat conduction in porous media. J Heat Transf 121:733–739

31. Hsu CT (2005) Dynamic modeling of convective heat transfer in porous media. In: Vafai K (ed) Handbook of porous media, 2nd ed. Taylor and Francis, London, pp. 39–80

32. Inamuro T, Yoshino M, Ogino F (1995) A non-slip boundary condition for lattice Boltzmann simulations. Phys Fluids 7:2928–2930

33. Imke U (2004) Porous media simplified simulation of single- and two-phase flow heat transfer in micro-channel heat exchangers. Chem Eng J 101:295–302

34. Jafari M, Farhadi M, Sedighi K, (2013) Pulsating flow effects on convection heat transfer in a corrugated channel: a LBM approach. Int Commun Heat Mass Transf 45:146–154

35. Kamiuto K, Yee SS (2005) Heat transfer correlations for open cellular porous materials. Int Comm Heat Mass Transf 32:947–953

36. Kao PH, Yang RJ (2007) Simulating oscillatory flows in Rayleigh-Bénard convection using the lattice Boltzmann method. Int J Heat Mass Transf 50:3315–3328

37. Kathare V, Davidson JH, Kulacki FA (2008) Natural convection in water saturated metal foam. Int J Heat Mass Transf 51:3794–3802

38. Kathare V, Kulacki FA, Davidson JH (2010) Buoyant convection in superposed metal foam and water layer. J Heat Transf 132:1–4

39. Khoei AR, Vahab M (2014) A numerical contact algorithm in saturated porous media with the extended finite element method. Comput Mech 54(5):1089–1110. doi:10.1007/s00466-014-1041-1

40. Lage JL (1998) The fundamental theory of flow through permeable media from Darcy to turbulence. In: Ingham DB, Pop I (eds) Transport phenomena in porous media. Elsevier, Oxford, pp. 1–30

41. Lai FH, Yang YT (2011) Lattice Boltzmann simulation of natural convection heat transfer of Al_2O_3/water nanofluids in a square enclosure. Int J Therm Sci 50:1930–1941

42. Lancellotta R (2002) Coupling between the evolution of a deformable porous medium and the motion of fluids in the connected porosity. In: Ehlers W, Bluhm J (eds) Porous media theory: experiments and numerical applications. Springer, Hamburg, pp. 199–225

43. Li CH, Peterson GP (2010) Experimental studies of natural convection heat transfer of Al2O3/DI water nanopartical suspensions (nanofluids). Adv Mech Eng 2010:742–739

44. Li ZH, Davidson JH, Mantell SC (2005) Numerical simulation of flow field and heat transfer of streamlined cylinders in cross flow. ASME J Heat Transf 128:564–570

45. Liu W, Davidson JH, Kulacki FA, Mantell SC (2003) Natural convection from a horizontal tube heat exchanger immersed in a tilted enclosure. J Sol Energy Eng 124:67–75

46. Liu W, Davidson JH, Kulacki FA (2005) Thermal characterization of prototypical ICS systems with immersed heat exchangers. ASME J Sol Energy Eng 127:21–28

47. Luan HB, Xu H, Chen L, Sun DL, He YL, Tao WQ (2011) Evaluation of the coupling shceme of FVM and LBM for fluid flows around complex geometries. Int J Heat Mass Transf 54:1975–1985

48. Mahjoob S, Vafai K (2008) A synthesis of fluid and thermal transport models for metal. Int J Heat Mass Transf 51:3701–3711

49. Mehrizi AA, Farhadi M, Afroozi HH, Sedighi K, Darz AAR (2012) Mixed convection heat tranfer in a ventilated cavity with hot obstacle: effect of nanofluid and outlet port location. Int Comm Heat Mass Transf 39:1000–1008

50. Muskat M (1937) The flow of homogeneous fluids through porous media. McGraw Hill, New York

51. Mohamad AA, Bennacer R, El-Ganaoui M (2010) Double dispersion, natural convection in an open end cavity simulation via lattice Boltzmann method. Int J Therm Sci 49:1944–1953

52. Mohamad AA, Kuzmin A (2010) A critical evaluation of force term in lattice Boltzmann method, natural convection problem. Int J Heat Mass Transf 53:990–996

53. Mondal B, Mishra SC (2009) Simulation of natural convection in the presence of volumetric radiation using the lattice Boltzmann method. Numer Heat Transf Part A 55:18–41

54. Nakayama A, Kuwahara F (2008) A general macroscopic turbulence model for flows in packed beds, channels, pipes, and rod bundles. J Fluids Eng 130:101–105

55. Nakayama A, Ando K, Yang C, Sano Y, Kuwahara F, Liu J (2009) A study on interstitial heat transfer in consolidated and unconsolidated porous media. Heat Mass Transf 45:1365–1372

56. Orszag S (1970) Analytical theories of turbulence. J Fluid Mech 41:363–386

57. Parmigiani A, Huber C, Bachmann O, Chopard B (2011) Pore scale mass and reactant transport in multiphase porous media flows. J Fluid Mech 686:40–76

58. Petrasch J, Meier F, Friess H, Steinfeld A (2008) Tomography based determination of permeability, Dupuit-Forchheimer coefficient, and interfacial heat transfer coefficient in reticulate porous ceramics. Int J Heat Fluid Flow 29:315–326

59. Peyret R, Taylor TD (1990) Computational methods for fluid flow. Springer, New York

60. Phanikumar MS, Mahajan RL (2002) Non-Darcy natural convection in high porosity metal foams. Int J Heat Mass Transf 45:3781–3793

61. Pouolikakos D (1986) Buoyancy-driven convection in a horizontal fluid layer extending over a porous substrate. Phys Fluids 29:3949–3757

62. Randive P, Dalal A (2013) Capillarity-induced resonance of blobs in a 3-D duct: lattice Boltzmann modelling. Int J Heat Mass Transf 65:635–648

63. Rong F, Guo ZL, Chai ZH, Shi BC (2010) A lattice Boltzmann model for axisymmetric thermal flows through porous media. Int J Heat Mass Transf 53:5519–5527

64. Riaz M (1977) Analytical solutions for single- and two- phase models of packed-bed thermal storage systems. J Heat Transf 99:489–492

65. Saito MB, Lemos MJS (2010) A macroscopic two-energy equation model for turbulent flow and heat transfer in highly porous media. Int J Heat Mass Transf 53:2424–2433

66. Sajjadi H, Gorji M, Kefayati GHR, Ganji DD (2012) Lattice Boltzman simulation of turbulent natural convection in tall enclosures using Cu/water nanofluid. Numer Heat Transf Part A 64:512–530

67. Shokouhmand H, Jam F, Salimpour MR (2009) Simulation of laminar flow and convective heat tranfer in conduits filled with porous media using Lattice Boltzmann method. Int Commun Heat Mass Transf 36:378–384

68. Sobieski W, Trykozko A (2011) Sensitivity aspects of Forchheimer's Approximation. Transp Porous Media 89:155–164

69. Su Y, Davidson JH (2007) Multi-zone porous enclosure model of thermal/fluid processes during discharge of an inclined rectangular storage vessel via an immersed heat exchanger. ASME J Sol Energy Eng 129:449–457

70. Su Y, Davidson JH (2007) Transient natural convection heat transfer correlations for tube bundles immersed in a thermal storage. ASME J Sol Energy Eng 129:210–214

71. Su Y, Davidson JH (2008) Discharge of thermal storage tanks via immersed baffled heat exchangers: numerical model of flow and temperature fields. ASME J Sol Energy Eng 130:021016-1–021016-7

72. Su Y, Davidson JH, Kulacki FA (2012) Numerical investigation of fluid flow and heat transfer of oscillating pipe flows. Int J Therm Sci 54:199–208

73. Su Y, Davidson JH, Kulacki FA (2013) Numerical study on mixed convection from a constant wall temperature circular cylinder in zero-mean velocity oscillating cooling flows. Int J Heat Fluid Flow 44:95–107

74. Su Y, Davidson JH, Kulacki FA (2013) A geometry factor for natural convection in open cell metal foam. Int J Heat Mass Transf 62:697–710

75. Su Y, Kulacki FA, Davidson JH (2014) Experimental and numerical investigations on a solar tracking concentrated photovoltaic-thermal system with a novel non-dimensional lattice Boltzmann method. Sol Energy 107:145–158

76. Su Y, Davidson JH (2014) A non-dimensional lattice Boltzmann method for direct and porous medium model simulations of 240-tube bundle heat exchangers in a thin rectangular enclosure. under review by International Journal Heat and Mass Transfer

77. Tarokh A, Mohamad AA, Jiang L (2013) Simulation of conjugate heat transfer using the lattice Boltzmann method. Numer Heat Transf Part A 63:159–178

78. Teitel M (2011) On the applicability of the Forchheimer equation in simulating flow through woven screens. Biosyst Eng 109:130–139

79. Vadasz P (2010) Heat flux dispersion in natural convection in porous medium. Int J Heat Mass Transf 53:3394–3404

80. Vafai K, Amiri A (1998) Non-Darcian effects in confined forced convective flows. In: Ingham D, Pop I (eds) Transport phenomena in porous media. Elsevier, Oxford, pp. 313–329

81. Vafai K, Kim SJ (1995) On the limitations of the Brinkman-Forchheimer-extended Darcy equation. Int J Heat Fluid Flow 16:11–15

82. Vafai K, Tien CL (1981) Boundary and inertia effects on flow and heat transfer in porous media. Int J Heat Mass Transf 108:195–203

83. Wang BX, Du JH, Peng XF (1998) Internal natural, forced and mixed convection in fluid-saturated porous medium. In: Ingham DB, Pop I (eds) Transport phenomena in porous media. Elsevier, Oxford, pp. 357–382

84. Wang M, Bejan A, (1987) Heat transfer correlation for Benard convection in a fluid saturated porous layer. Int Comm Heat Mass Transf 14:617–626

85. Wade A (2010) Natural convection in water saturated metal foam with a superposed fluid layer, Masters Thesis, University Minnesota, August 2010

86. Wakao N, Kaguei S (1982) Heat and mass transfer in packed beds. Gordon and Breach Science, New York

87. Ward JC (1964) Turbulent flow in porous medium. ASCE J Hydraul Div 90, HY 5:1–12

88. Whitaker S (1977) Simultaneous heat, mass and momentum transfer in porous media: a theory of drying. In: Hartnett JP, Irvine TF, Jr (eds) Advances in heat transfer, vol. 13. Academic, New York, pp. 119–203

89. White FM (1991) Viscous fluid flow. McGraw-Hill, New York, pp. 181–184

90. Xu RN, Huang YL, Jiang PX, Wang BX (2012) Internal heat transfer coefficients in microporous media with rarefaction effects. SCIENCE CHINA Tech Sci 55:2869–2876

91. Yang C, Nakayama A (2010) A synthesis of tortuosity and dispersion in effective thermal conductivity of porous media. Int J Heat Mass Transf 53:3222–3230

92. Yokokawa M, Itakura K, Uno A, Ishihara T, Kaneda Y (2002) 16.4-Tflops direct numerical simulation of turbulence by a Fourier spectral method on the Earth simulator. Proceedings of the ACM/IEEE Conference on Supercomputing, Baltimore
93. Zhang DX, Hang RY, Hen SY, Soill WE (2000) Pore scale study of flow in porous media: scale dependency, REV, and statistical REV. Geophys Res Lett 27:1195–1198
94. Zhao CY, Lu TJ, Hodson HP (2005) Natural convection in metal foams with open cells. Int J Heat Mass Transf 48:2452–2463
95. Zhao CY, Dai LN, Tang GH, Qu ZG, Li ZY (2010) Numerical study of natural convection in porous media (metals) using Lattice Boltzmann Method(LBM). Int J Heat Mass Transf 31:925–934
96. Zhao CY (2012) Review on thermal transport in high porosity cellular metal foams with open cells. Int J Heat Mass Transf 55:3618–3632
97. Zhukauskas A (1972) Heat transfer from tubes in cross flow. In: Hartnett JP, Irvine TF, Jr (eds) Advances in heat transfer, vol 8. Academic, New York, pp. 21–28